Self-motioned Atomic
Dialectical Materialist Logic

Self-motioned Atomic Dialectical Materialist Logic

SCIENTIFIC PHILOSOPHY
AND NATURAL SCIENCES
MUST MAINTAIN ETERNAL
RELATIONSHIP TO MAKE SENCE

———

Hossein Dallalbashi

FIRST EDITION

Copyright   1979 by Hossein Dallalbashi

Published by Vantage Press, Inc.
516 West 34th Street, New York, New York 10001

Printed in the United States of America
Standard Book Number 533-03659-3
ISBN: 1530103924
ISBN 13: 9781530103928

Library of Congress Catalog Card No.: 78-51742

Table of Contents

Introduction· ·ix

Chapter 1   Universe Is Made of Diversified, but Related Matter. · · · · · · 1
           1.-Materiality and Objectivity of Matter· · · · · · · · · · · · · 5
           2-The Concept of Matter Disregards Individual
           Distinctions.· 5
           3.-Solid, Liquid, and Gaseous Entities As the Most
           General Forms of Matter · 6

Chapter 2   The Concept of Matter Is Too Broad, Too Block-Like
           to Represent Material Universe;   There Is a Need to
           Introduce a Smaller Building-Block (Atom).· · · · · · · · · · · · 7
           1.-Atomic Structure As the Smallest Building-Block;
           As a Unit of Atomic Universe · · · · · · · · · · · · · · · · · · · 8
           2.-Atoms and Motion· 10
           3.-Motions Are Primarily inherent in All Atoms· · · · · · 10
           4.-Motions As General Mode of Existence of Atoms· · · 11

Chapter 3   Forms of Motion · 14
           1.-Nature Itself Is Constantly Overlapping.
           Therefore, Separate Treatment of Motions Is Unscientific
           and Bad Philosophy.· 15

2- Self-Motion As the Basis of Specific and Internal
Forms of Existence of Atoms (Self-Motioned Atomism) · · 16
3.-Atomic Motion Is Not Primarily Externally Caused.
It Is Self-Contained (Self-Motioned Atoms). · · · · · · · · · · 18

Chapter 4   Some General Observations and Conclusions on the
Structure and Behavior of Self-Motioned Atoms · · · · · · · · 20
1.-Interrelationship of Self-Motioned Atoms · · · · · · · · 21
2.-Self-Motioned Atoms Have Varying Combining
Abilities, Forming More Complex Self-Motioned
Entities. · 21
3.-Self-Motioned Atom Is Dialectical in Nature. · · · · · 23
4.-Evolution of Self-Motioned Atoms · · · · · · · · · · · · · · 24

Chapter 5   Self-Motioned Atomic Universe Is Law-Governed. · · · · · · · 28
1.-Laws As Involved, Complicated Interaction and
Relationships of Self-Motioned Atomic, Molecular,
and Macro-Body Entities · 29
2.--Particularity of Unfoldment of Law · · · · · · · · · · · · 30
3.-Evolution of Self-Motioned Atoms and Creative
Unfolding of Law · 31
4 -Operation of Laws Depends on the Complexity of
Different Parts of Self-Motioned Atomic Universe · · · · · 31
5.-Laws Are Not Immutable; They Come Into Being
Creatively During the Process of Evolution. · · · · · · · · · 33

Chapter 6   The Concept of Time Tied Up to General Motion and
Movement · 35
1-Time Is Not Simply Reduced to Simple Mechanical
Motion and Movement; But It Is Intimately Associated
With Evolution of Self-Motioned Atomic Universe. · · · · · 36
2.-Particularity and Generality of Time · · · · · · · · · · · · 37

     3.-Time Is Born Out of, Lives Through, and Is
Determined by Self-Motioned, Inter-Flowing
Atomic Universe. · 38
     4.-Space Is Not an "Empty Container" Separate
From Its Material Content, the Self-Motioned
Atomic Universe. · 39
     5.--Particularity and Generality of Space · · · · · · · · · · · 39

Chapter 7 Laws of Evolutionary Development of Self-Motioned
     Atomic Universe· 41
     I.-The Law of Unified and Opposing
Interpenetration of Self-Motioned Atoms · · · · · · · · · · 42
     2.-Universality and Relativity of Unity and Opposing
Tendencies of Self-Motioned Atoms· · · · · · · · · · · · · · 44
     3- The Law of Qualitative Transformation of Self-
Motioned Atoms · 47
     4.-Particularity of Qualitative Transformation of
Various Parts of Self-Motioned Atomic Universe· · · · · · 50
     5.-The Law of Negation of Negation of
Self-Motioned Atoms · 53

Chapter 8 Causality-Cause and Effect · 58
     I.-Self-Motioned Atom, Being *Primarily* Self-Caused,
Secondarily Engages in Mutually Determined
Causations. · 58
     2.-The Relationship Between Self-Motioned
Atoms and the Rest of the Material Universe.· · · · · · · · 58

Chapter 9 Truth · 66
     1.-Simple Reflection and Correspondence of Material
Universe is Not Enough to Establish Truth: An
Evolutionary Understanding Is Needed. · · · · · · · · · · · 68

2 -Evolutionary Self-Motioned Atomic Universe As
the Basis of Objective Truth · 69
3.-Stationary, Non-changing, Non-evolving Material
Universe and the Problems of Establishing Truths · · · · 69
4.-Self-Motioned Atomic Universe Is Not Fixed,
' Stationary; It Is Constantly Changing and Evolving. · · · 70
5.-Relativity of Existence of Self-Motioned Atomic
Universe · 72
6 -Relativity vs. Absoluteness of Truth · · · · · · · · · · · · · 78

Introduction

———

HUMAN HISTORY, AT LEAST IN the last two thousand years, has attempted to understand the material universe, including its highest product, human beings. This called for trying *to* comprehend the *material causes* of everything around us. Those who advocated that material universe can only be investigated by trying to under stand its material causes were called materialist philosophers and scientists. We therefore had the term materialism, philosophical materialism, brought about, expressing this point of view. Ancient philosophers wanted to know what material universe was made of in order to understand it. Daily observations showed that every thing that human beings experience on a daily basis changes. So, at this time, the problem was twofold: (1) To find out the material causes of the material universe, and (2) why it keeps changing. It was not until recent times, when it was discovered by scientists that material universe not only changes in parts, but rather *things evolve and develop and become something else.* It was also discovered that human societies too, change, evolve, and develop. What could then be the material causes of the evolution of human society.

At this juncture, the problem becomes more complicated. We have several issues to deal with: first, we have to investigate what our material universe is made of. Second, why does every thing keep changing? Third, what causes the material universe to evolve and develop, becoming something else in parts, as well as in its " entirety", now that we know everything is evolving, never staying the same, what are the laws in accordance with which everything evolves and develops. There have been different disciplines, such as physics, chemistry, and biology, not to mention social sciences developed in order to provide answers to these basic questions. These sciences themselves have evolved in the last two thousand years of human society. The answers, therefore, provided by these sciences have also varied. But, in the twentieth century, these three basic sciences are gradually being forced to explain everything in their existence, change, movement, and evolution. In short, gradually, they are becoming evolutionary sciences. Corresponding to every stage of development of natural sciences, within the context of a given social development, there arose various philosophers attempting to explain the basis of life. There also arose a philosophy of materialism which attempted to put all these sciences together and tell us what we are doing, and where we are going. So from ancient times up to the present, the philosophy of materialism has had a very intimate relationship with natural sciences, only to become a love affair at a later time.

Natural sciences did not always deal with the problems of evolution. So we can safely maintain that up to the latter part of the nineteenth century, natural sciences were of non-evolutionary character. Therefore, since the philosophy of materialism always relied upon natural science to explain different phenomena, then up to that time, it too expressed itself on a. non-evolutionary basis. And since the science of mechanics was one of the most developed of natural sciences in the nineteenth century, it influenced the materialist philosophy of that time. It was, therefore, called mechanical materialism. Mechanical materialism did not enjoy the support of modern physics, chemistry, or biology, only to be reunited with them at a later time.

That means every time there is one or a series of discoveries in natural sciences, physics, chemistry, and biology, then this philosophy of materialism has to *revise itself,* in order to. be worthy of its intimacy, therefore love affair, with natural science. Since all natural sciences are being revolutionized and established on evolutionary foundation, philosophy of materialism too must reestablish itself on an evolutionary basis. Well, the initial act was taken by Marx and Engels to convert the philosophy of materialism to that of an evolutionary one, This took place around one hundred years ago. They called their evolutionary materialism "dialectical materialism." In this, they not only explained the processes of evolution in nature and their material causes, but they also explained the material reasons of evolution of human society to a higher level.

At this point my contention is an attempt to show that the very intimate love affair between dialectical materialism and natural sciences, while being firmly rooted in human society, is one of permanent nature and that for philosophy to maintain its relevancy, it must constantly pay attention to the new discoveries of natural **sciences.**

To support this argument, I shall present quotations from Engels, Lenin, and two of the most prominent physicists of modern times, de Broglie and Albert Einstein.

Here Engels points out that new discoveries in natural sciences *must result in a change of our materialist philosophy:*

But just as idealism underwent a series of development, so also did materialism. With each epoch-making discovery, even in the sphere of *natural science* it *has to change its form* ... The materialism of the last century was predominantly mechanical because, at that time, of all natural sciences, mechanics and, indeed only mechanics of solid bodies-celestial, terrestrial, in short, the mechanics of gravity, bad come to any definite close. Chemistry, at that time, existed only in its infantile philogistic form. Biology still lay in swaddling clothes; vegetable and animal organisms had been only roughly examined and were explained as the result of mechanical causes.[1]

In Engels's time, the latter part of the nineteenth century, there had been three discoveries related to natural sciences. They were: (1) the discovery of cells in biological organisms, (2) the idea that motion transforms, in physics, and (3) Darwin's evolutionary theory, indicating that man was not created; he evolved out of lower forms of animals.

Marx and Engels critically incorporated these discoveries in what was called mechanical materialist philosophy, to be developed into an evolutionary materialism.

Engels was very excited about this. He regretted the fact that one of the greatest materialist philosophers of all time, Feuerbach, did not have access *to* these discoveries in order to make use of them. Here is what he had to say about it:

It is true that Feuerbach had lived to see all three of the decisive discoveries-that of the cell, the transformation of energy and the theory of evolution named after Darwin. But how could the lonely philosopher, living in rural solitude, be able sufficiently to follow scientific developments in order to appreciate their full value, discoveries which scientists themselves at that time, either contested or did not adequately know how to make use of. The blame for this falls solely upon the wretched conditions in Germany, in consequence of which cobweb-spinning eclectic flea-crackers had taken possession of the chairs of philosophy, while Feuerbach, who towered above them all, had to rusticate and grow sour in a little village. It is, therefore, not Feuerbach's fault that the historical conception of nature, which had now become possible, which removed all the one-sidedness of French materialism, remained inaccessible to him. Secondly, Feuerbach is correct in asserting *that, exclusively, natural-scientific materialism was indeed "the foundation of the edifice of human society. Knowledge, but ... not ... the building itself."*[2]

Engels goes on to add:

Thanks to these great discoveries and the other immense advances in natural sciences, we have now arrived at the point where we can demonstrate, as a whole, the inter-connection between the processes in nature not only in particular spheres, but also in the inter-connection of these particular systems, and so can present in an approximately systematic form a comprehensive view of the inter-connection in *nature by means of facts provided by these empirical sciences.*

Lenin, in his book, *Materialism and Emperio-Criticism,* employing a quotation from Engels testifies to the same attitude:

Engels says explicitly that with each epoch-making discovery even in the sphere of natural science (not to speak of history of mankind)," *materialism has to change its form.* Hence, *a revision of the natural-philosophical propositions is not only **not revisionism"** in the accepted meaning of the term, but on the contrary, *is an essential requirement of Marxism.* [4]

Engels goes on to express his scientific attitude toward the relationship of dialectical materialism and natural sciences and the fact that the former *must constantly redeem itself* by maintaining a close· contact, of a *blood-line nature,* with the latter. Engels holds:

This modern materialism, (dialectical materialism) the negation of negation, is not mere establishment of the old (mechanical materialism), but adds to the permanent foundations of this old materialism the whole thought-content of *two thousand years of development of philosophy and natural science,* as well as the history of these two thousand years. It (dialectical materialism) is no longer a philosophy at all, but simply a world outlook which bas to *establish its validity and be applied not in a science of sciences (i.e., a philosophical system) standing apart, but* in *the positive (natural) sciences.* Philosophy (dialectical materialism) is, therefore, "sublated" here, that is, "both overcome and preserved," overcome as regards its form, and preserved as regards its real content. [Engels goes on to add:] In both cases modern materialism is essentially dialectical, *and no longer needs any philosophy above the other sciences.*•

And lastly, with a profound degree of modesty and personal humility, which can only be Engels's way of doing things, realizing the significance of dialectical materialism as a scientific logic, being quite cognizant of its level of development at the time, and the fact that it had to further evolve, he even goes as far as saying that his dialectical materialism, going through new discoveries of natural sciences "may possibly make my (his) Work, to a great extent, or even all together, superfluous." Engels goes on to conclude that: "And this materialistic dialectics *which for years has been our best working tool and our sharpest* weapon was, remarkably enough, discovered not only by us (Marx and Engels), but also independently of us and even of Hegel by a German worker, Joseph Dietzgen."•

Yet the advance of theoretical natural science may possibly make my work to a great extent or even altogether superfluous. For the revolution, which is being forced on theoretical natural science by the mere need to set in order the purely empirical discoveries, great masses of which have been piled up, is of such a kind that it must bring the dialectical character of natural processes more and more to the consciousness even of those empiricists who are most opposed to it!

Basing my judgment upon the materials that are available in English in the U.S., I think that Engels's tradition is not being carried out and therefore dialectical materialism is not maintaining a sufficient contact with natural sciences; physics, chemistry, and biology, and other offspring sciences derived from these fundamental natural sciences, let alone having a juicy love affair relationship with them.

The separation between dialectical materialism and natural science seems to be widening. Instead, dialectical materialism has been forced to establish a closer contact with politics and has, there fore, become *highly politicized*. Dialectical materialism, in Marx and Engels tradition, was highly ingrained and founded on natural sciences. The closer this logic would get to natural sciences, the more useful, effective and powerful it would become as a superior form of reasoning; the wider the gap between the two, the shallower the outcome. Nowadays, the attempt to see dialectical materialism in terms of natural sciences and their development is at worst abandoned and at best taken only passively. Instead, a collection of political slogans, positions and statements are presented in defense of definition and direction of dialectical materialism. How did that come about? Immediately after the following quotations, I will deal with this issue.

The harmfulness of this separation between philosophy and natural sciences was well recognized by two of the most prominent natural scientists of modern time, Prince Louis de Broglie, who created the wave theory of matter, and Albert Einstein, the author of "theory of relativity."

Here is what de Broglie writes:

In the nineteenth century there came into being a separation between scientists and philosophers. The scientists looked with a certain suspicion upon the philosophical speculations, which appeared to them, too frequently, to lack precise formulation · and to attack vain, insoluble problems. The philosophers, in turn, were no longer interested in the special sciences (natural sciences) because their results seemed too narrow. This *separation, however, has been harmful to both philosophers and* **scientists.**

Albert Einstein reinforces this argument by the following statement:

I can say, with great certainty, that the ablest students, whom I met as a teacher, were deeply interested in theory of knowledge (philosophy), I mean by "ablest students," those who excelled not only· in skill, but in independence of judgment .. .[9]

Here Einstein's reference to "students" means students of natural sciences, interested in philosophy.

Here, both de Broglie and Albert Einstein are pointing out to the unfortunate separation of philosophy, in general, and natural science. My concern is the widening gap between dialectical ma materialist philosophy and natural sciences.

And finally, Karl Marx, seeing the real source of development of philosophy and natural sciences within the context of social development, and more specifically, the requirements of industry, being aware of their transitory, artificial separation, concludes that there is a gradual unification of philosophy (dialectical materialism) and natural sciences only to reach a blossoming stage in a communist society.

"Industry," says Marx, "is the actual, historical relationship **of nature, and, therefore, of natural science to man.**"[10]

The natural sciences have developed an enormous activity and have accumulated an ever growing mass of material. philosophy, however, has remained just as alien to them as they remain to philosophy. Their momentary unity was only a chimerical illusion. The will was there but the means were lacking."

Natural science has invaded and transformed human life all the more practically through the medium of industry and has prepared human emancipation, although its immediate effect had to be the furthering of the dehumanization of **man.**[12]

Natural science, [Marx concludes,] will in time incorporate into itself the science of man, just as the science of man will incorporate into itself natural science: *There will be one* **science.13**

To respond to the social cause of this separation, we have to summarily follow, at least in very general terms, the development of capitalist society from the time when it was born out of the womb of the feudal system.

At that time the feudal society basically consisted of two social classes; the peasantry, who worked on the land and lived based upon their own labor, and the landowners, or the aristocracy, who lived off the labor of the peasantry. The emergence of capitalist society, based upon advancement of productive power of the society, shook the feudal foundation, where the aristocracy and peasantry, as two main social classes, were gradually forced to disappear, only to be replaced by two modern social classes; the industrial working class and the capitalists, or owners of production facilities, factories, and agricultural and industrial corporations.

The working class came out of the farm laborers by the re requirement and development of industries. It had very little education, skill, and scientific knowledge. The bulk of the working class was basically physical, doing physical work. The capitalists came out the more aggressive section of the aristocracy. It had very little power in the beginning, but it had to struggle to establish itself in all walks of life. Today it has not only established its control over the most important activity of human society (economic), but also controls the educational systems, judicial system, army, and police. You name it, they control it.

The working class too has gone through a very complicated process of development. Again responding to the requirements of industries, it had to acquire more skill, knowledge, and education, in order to become more productive. The logic of capitalist soci ety and its corresponding competitive character constantly called for introduction of more efficient forms of production. The urge for profit made the capitalist use more machinery than relying upon doing things with hands (manual orientation). There gradually de veloped a process of mechanization of the economy. This meant that mental labor, those who could make productive machines, would gradually play a more important job in terms of advancing productive ability of the society and, therefore, speeding up the process of social transition. And if an industry relied more on physical labor, it had to have meant that the capitalists profited more by the use of cheap physical labor with minimal employment of productive machinery used in that industry, than in putting a greater emphasis on employment of mental labor or the use of highly mechanized productive facilities. Therefore, predominant employment of physical labor characterizes the early part of capitalist development, whereas the development of society resulting in introduction of more machinery in production processes, characterizes the more modem form of the working class with mental productive power. We might say that the working class is becoming more "mentalized" as opposed to its earlier form of existence which was basically more "physicalized."

Today there is not a single commodity, ranging from toilet tissues to spaceships exploring other planets, produced in highly industrialized capitalist societies, whose production does not need the use of highly efficient machinery.

In order to produce these very advanced and sophisticated productive facilities or consumer goods, the mental workers had to acquire very advanced technical and scientific knowledge. That meant that only menial jobs, mostly service industry, could still rely upon physical workers. The rest of the society needed more and more mental workers to respond to its needs and requirements. But we observe that the majority, engaged in productive activity in a highly industrialized society, are made up of physical and mental workers and that the capitalists (owners of productive facilities) only constitute the "minority." In the U.S., the ratio is 20 percent to 80 percent, respectively.

The capitalists from the early days had to organize political institutions (parties) that guaranteed their overall interests. In the United States, the Democratic and the Republican parties, with slight ideological differences, both stand for the interests of the in industrialists, the capitalists. You might say that minority social class made sure that they were going to rule the majority through these political institutions and that, above all, they concerned themselves with maintaining the existing economic order.

But what happened to the majority-the working class? How would it express its point of view, ensure its interests on an economic and political basis. Did it form its own political party to take care of this problem as the capitalists had done for themselves? In other countries, the working class formed its own political parties under the names Socialist and Communist parties. What about the working class in the U.S.? ·Despite the fact that there is a Communist Party and other working class splinter groups, the working class is told that the Democratic party is the one they belong to. It is the party that is concerned about working-class interests, as the story goes. You have never heard such a big lie in your life! The point being, that the working class in America was never allowed to form its own independent political institution to represent its interests. This honor can. only go to the Communist Party of the U.S. of A., and other genuine working-class organizations that have always been the victims of systematic harassment and attempts to discredit them by the Democratic and Republican parties and their puppet instruments of oppression, the F.B.I. and the C.I.A. The existence, purposes, and aims of these working-class organizations have been relentlessly falsified and their development checked by the entire capitalist conditioning educational system and culture.

But the working class, being the majority, is interested in transforming the society into majority rule. That calls for the transition of the society to a more humanized form of human society, namely, socialism. Only the working class as a whole is capable of this tremendous job.

Marxism was born to represent the interests of the working class· and explain its importance in transforming the society. It provided a world outlook, a scientific body of thought that explained the process of social evolution, with the working class as the main performer, while the freelance intellectuals do the Olympic gymnastics.

To give direction to this social transformation, Marxist parties were also born. This was in the latter part of the nineteenth century. At that time, the working class was mainly "physical"; there were very few mental workers. Intellectuals from other social classes, inspired by humanitarian feelings, joined the working-class Marxist parties to provide leadership and direction to them.

Marx, Engels, Lenin, Mao, Fidel Castro, to mention a few, all joined the working-class cause. *They were not physical workers themselves.*

Almost any Marxist party can point out to many of its leaders who were not physical workers themselves, but they joined the working-class parties.

My concern at this time is to explore some of the problems related to Marxist parties and the problems of social transition in highly industrialized capitalist societies;

Marxist parties have always been simultaneously helped by and distrustful of intellectuals, in order to achieve an easier social transition. This duality has always existed. On the one hand, the working class was developing its own culture and had to rely upon intellectuals to verbalize this culture as was the case with Marx, Engels, and Lenin with their theoretical contributions in developing a materialist world outlook. On the other hand, they were fearful of having certain intellectuals taking over the leadership and theoretical position, bringing other alien philosophical tendencies in Marxist parties, which were basically anti-working class in nature. As a matter of fact, this fear is not entirely un justified, and there have been these alien encroachments upon Marxism from time to time. Sometimes this fear has been un necessarily exaggerated. The workers Communist movement is suffering from alien philosophical encroachments that are living off the scientificity of Marxism.

This fear of intellectuals with alien philosophy and the development of the working class is my concern at this time.

The physical workers needed outside intellectuals, who agreed with working-class positions, to express what the physical workers themselves could not do, *verbalize* the materialist outlook and the working-class interests on a scientific basis. Therefore, they had to be sure of whom they do or do not include in the workers' movement.

Today the working class has acquired the greatest degree of technical and scientific knowledge, capable of sending exploratory space vehicles to other planets. Its scientific productive powers are almost unlimited. This *scientific knowledge* constitutes the *working-class culture* being *born within the womb* of the capitalist society.

Marxist parties, especially in highly industrialized countries, need not borrow intellectuals anymore to explain their point of view. The working class itself has developed its own scientists and intellectuals; they are mental workers,· engaged· in research and production; they are physicists, chemists, biologists, technologists, engineers, and technicians, just to mention a few. You name it and the working class has developed it. Modem production could not move forward even an inch without these mental workers. The production no longer starts at the factory level. Scientific research workers are just as productive as others involved in actual production. Some Marxist ·parties ignore the best part that the working class has developed, mental workers. Why?

One hundred years ago Marx and Engels provided a scientific, evolutionary logic, which they called dialectical materialism. It represented the best of working-class culture. It was supposed to have been further developed by the working class as they acquired more scientific knowledge. Only the organized mental labor could have done that. But if the mental labor is despised, excluded from Marxist parties, which are remaining "physical" in orientation, then how can that happen? Who would be interested in developing that outlook? The physical workers cannot still expect the sym sympathizing, outside intellectuals, with their sketchy, half-ass knowledge to come to their help. Those days are gone forever.

In the highly industrialized capitalist countries, mental and physical workers must unite and employ the best scientific knowledge of the former and the diversified and rich experience of the latter in developing a materialist outlook, in order to expedite the social transition.

But this has not happened. The mental workers are not developing dialectical materialism nor do the physical workers have their contributions recognized on a social level. Instead; physical workers are still following the borrowed intellectuals to explain their position and "save them," at best, or dovetailing bourgeois parties and becoming totally dissolved and subordinated,   at worst.

These borrowed intellectuals are primarily political. They have politicized dialectical materialism, engaging in shallow political sloganeering.

At times, these borrowed intellectuals, being detached from the daily problems of the working class, put forth ideas, "theories," that the working class does not identify with nor is even able to digest.

They would either want to lead the workers movement to the right, which would result in subordination of the working class to bourgeois parties, or to ultra-left positions, which can only mislead the working class into an impasse. All that happens, because the working class does not rely upon itself and its own best creation, the mental, scientific workers. That is why we have people like Mao, "the philosopher of the East" writing a recipe as to how every society, including American, should transform. How naive and stupid! It looks more like John Wayne appearing as a hero and saving America from extinction. It can only be true in a Hollywood movie. The working class does not need heroes. They have to rely on their own experience and scientific knowledge.

And what happens when the working class does not rely upon these borrowed intellectuals, instead of paying attention to its own culture, is that proletarian culture is being reduced to watered down, lopsided, and inadequate political and economical issues. The physical working class is not capable of paying attention to what Engels called the foundation of human society; physics, chemistry, biology, and so forth. The working class culture, therefore, remains lopsided. What happens is that a fantastic degree of scientific knowledge that could be used by mental scientific workers to further develop working-class culture, a materialist educational system, dialectical materialism, is instead being used by the capitalists to "improve" their oppressive system and justify it. Not *to* mention the fact that the mental workers, being alienated from its own organic part, physical workers are co-opted to the other side to be used and misused towards maintaining one of the most oppressive, outmoded social systems in modem times, which bas the greatest disrespect toward people, namely, capitalism.

Today capitalism does not reduce itself to sociopolitical and economic arguments against the working class; it misuses natural science to "prove" that capitalism is to stay with us forever, with the capitalist tyrants in the leadership position. Every discovery in natural science proves beyond the shadow of a doubt the validity of the evolutionary logic, dialectical materialism. *Then why should the working-class voluntarily impose rigid limitations upon itself by reducing its arguments against capitalism to political and eco nomic issues.* Our disagreement goes beyond that. We have differ ences with the capitalist outlook in almost any subject you can think of. The greatest weapon, which is a by-product of working class culture, is the materialist outlook.

Marxist parties, especially in the West, uniting with and relying on the mental section of the working class, must develop party physicists, chemists, and biologists on a dialectical materialist basis. This would deliver the Marxist parties from the lopsidedness they suffer, in order to be better able to compete with the capitalist educational system and finally expose the shortcomings of capitalist systems on all levels. I remain convinced that should Marxist parties of the West continue excluding mental labor from their leadership and rank-and-file positions, it is highly doubtful that physical workers alone can successfully prepare for a social transformation to socialism in the Marxist sense of the term.

Some Marxist parties act as if the case for development of Marxism, especially dialectical materialist logic, bas been closed and that Marxism can only teach and not learn. I strongly disagree with this point of view.

Granted, once in a while, we hear some noise that this or that individual "developed Marxism." It so happens that when we study that "development," we see that the alleged development is nothing but ·a regression of dialectical materialist thinking to its more primitive form of existence or a naive attempt to pass idealist philosophy for dialectical materialism.

A case in point is the "development" of Marxism by Mao, who more often engages in professing religiously oriented, divine rev elations than seriously dealing with the science of dialectical materialism.

As for my work, nothing has absolute and eternal validity. It so happens *that at this time,* I remain convinced as to the scientific validity of what I have done in this book. However, should I be come convinced by criticisms of my work by Marxist physicists, chemists, biologists, and social scientists that revisions in parts or of the entire book have to be made, I shall be only too glad to do so.

In the remaining pages of the Introduction I have, in most general terms, explained what one can expect in reading this book, only to be expounded more comprehensively in the book itself.

Greek philosophers, Democritus, Thales, Heraclitus, Anaximanese, and Anaximander, were some of the most prominent materialist philosophers of ancient times, who recognized that the material universe is in constant change and evolution. They attributed *the causes of change* to different opposing forces found in nature, such as, cold, warm, fire, and water. They attempted to find reasoning that explained the processes of change based upon these opposing forces.

This seemingly simplistic explanation was the beginning of a more profound explanation to be developed in the nineteenth century by Marx and Engels, who discovered that the material universe, from the smallest particles (sub-atomic and atomic) to the largest planets, develops based upon opposing forces found in nature, i.e., positive and negative electric and magnetic charges in atomic structure. Further observations made them convinced that these opposing forces (tendencies) can only exist within the context of "an organized unit." The process of interaction of these opposing forces was termed "contradiction." That simply meant that the opposing forces, while maintaining a unity, were inter penetrating with one another and that the processes of interaction and interpenetration result in further evolution of that object. The book deals with the problem more comprehensively.

I have found the term "contradiction" to be very misleading, creating confusion for some students of Marxism. Besides, bour geois theoreticians have attached phony moral and ethical values, which are their own fabrications, to the concept of "contradiction," in order to create more confusion. Engels, in his writings, more than once equates "contradiction" with the process of motion. Based upon his teachings of "contradiction," I therefore have re placed the concept of "contradiction" with the term, "process of **self-motion.**"

The following quotation from Engels, in my opinion, justifies my act:

True, so long as we consider things as at rest and lifeless, each one by itself, alongside and after each other, we do not run up against any contradictions in them. We find certain qualities which are partly common to, partly different from, and even contradictory to each other, but which in the last-mentioned case are distributed among different objects and, therefore, contain no "contradiction" within. Inside the limits of this sphere of observation we can get along on the basis of the usual metaphysical mode of thought. But the position is quite different as soon as we *consider things in their motion, their change, their life, their reciprocal influence on one another.* Then we immediately become involved in contradiction. *Motion, itself, is a contradiction.* Even simple mechanical change of position can only come about through a body being at one and the same moment of time both in one place and in another, being in one and the same place and also not in it. *And the continuous origination and simultaneous solution is precisely what motion is.*[1]*

Other Marxists talk about "object and phenomena," the development and process of self-movement associated with them, just as Darwin explained the process of biological evolution in terms of the relationship between biological organisms and their environ ment (natural selection). This is an attempt to see evolutionary process on a macro-body level. But then, yon cannot treat organisms and environment as two completely separate, finished units, without internal-structure interacting. Both the environment and organisms are made of an infinite number of individual self motioned atoms, molecules, genes, cells, organs, every one of which has interrelated individual existence and distinctions, which, when combined, make up bigger bodies. It therefore makes it necessary to investigate the existence, composition, and behavior of the smallest component parts of the material universe.

So, I have, in my own mind, followed Engels's teachings and, therefore, traced the process of self-motion, and self-movement of the individual atom, molecule, and macro-body and their rela tionships with the rest of the material universe, just as modern physics (quantum mechanics) deals with objects and phenomena on a sub-atomic and atomic basis, and biology deals with biologi cal organisms on a molecular level. My attempt is "to go further" and explore the process of evolution on a micro-level, and having done so, establish a relationship between micro-and macro-entities. There is so much that dialectical materialism can teach quantum mechanics and molecular biology and so much that it can learn from them. *The influence is of mutually determining nature.*

My work is a modest attempt to explain dialectical materialism in a non-technical, popular basis; yet it is not quite a traditional approach to explaining dialectical materialism. It represents the beginning of a long overdue attempt to *revise dialectical material ism in order to accommodate the new advances in natural science and further social development,* just as Engels wanted it to be. This in the final analysis, is the collective job of scientific workers.

Some self-styled Marxists act as if natural sciences have not developed since Engels's time. There are those who even deny there is a relationship between Marxism and natural sciences. What an understatement! At that time Engels felt that materialism was given new blood by the discoveries of the cell, transformation of energy,. and the Darwinian evolutionary theory. At that time quantum mechanics, nuclear physics, molecular biology, and the beautiful science of genetics were either non-existent at worst, or they were in their embryonic stages at best. Today these sciences are well developed and capable of offering wonders.

If studied critically and incorporated into dialectical materialist logic, they truly revolutionize dialectical materialism be yond recognition. But what happens is that people like Mao, who may have undoubtedly been instrumental in the Chinese Revolution, having exhausted their meaningfulness and relevancy, and who probably never saw a book on theoretical natural sciences let alone having studied it in the Engels tradition, are claiming to have "developed" dialectical materialism. This "development" is nothing but the juggling of political slogans, taken seriously by politically naive persons, only to become totally disillusioned after they learn the ABC of Marx and Engels dialectical materialism. He and his kind, of which unfortunately we have too many in the workers' Communist movement, are nothing but an imposition upon the working class and its historical role.

I am sure, "absolutely sure"-and this is the only "absolute" that I admit to-that Chinese society will soon by-pass Mao and many others of the same mentality, as the process has already begun. That does not mean that Maoist dogma would not leave a mark and therefore, a bad influence upon the Chinese social development, making it more difficult to evolve its transition smoothly.

Having attempted to explore dialectical materialism from the position of self-motioned atoms and molecules, I, therefore, have slightly modified the term dialectical materialism to accommodate the· elaboration of the concept of "self-motion," which I have elaborated upon. Hence, I have called it "self-motioned, atomic dialectical materialism." It is a little mouthful and I admit that. But I had to do it so that a distinction could be made between my understanding of dialectical materialism and those of others.

Of special importance are the (1) introduction and elaboration of the process of self-motion or self-movement as found in nature on atomic, molecular, and macro-body levels, (2) elaboration of laws, claiming that *they are being born in the process of evolution and that their unfoldment correspondingly changes, be coming more complex.* I have, therefore, concluded that laws are not *"immutable"* and have, whereby, put them on a more dynamic foundation, (3) elaboration of causality, with emphasis on *mutual determination of a self-motioned atomic universe,* which is, in my opinion, much closer to Engels's teachings than currently under stood by some Marxists, (4) elaboration of the concept of truth showing relativity of material existence and, based upon this, establishing relativity of truth as opposed to "absoluteness of truth" that some Marxists maintain.

In analysis of dialectical materialist understanding of truth, I have completely disavowed myself from any concept of "absolute truth," especially if it is to be found in the most alien and strangely unexpected places, in dialectical materialist logic. "Absolute truth" does not go with my state of mind and psychology, even if it had any scientific validity, which it undoubtedly does not!

1. F. Engels, *Ludwig Feuerbach and the Out come of Classical German Philosophy* (New York: International Publishers, 1941), p. 26.

2. Engels, *German Philosophy,* pp. 28 29.

3. Ibid., pp. 4647.

4. N. Lenin, *Materialism and Empirio-Criticism* (Moscow: Progressive Pub., 1941), p. 251. ..

5. F. Engels, *Anti·Duhring* (Moscow Progressive Publishers, 1968), p. 23.

8. Phillipp Frank_,.*Science of Philosophy* (Englewood Cliffs, N.J.: Prentice Hall, Inc., 1957), p. ll.

9. Ibid., p. 11.

10. K. Marx, quoted by *New Left Review*, 82, p. 64, Source, Economic and Philosophic Manuscript II. Ibid., p. *65.*

12. Ibid.

13. Ibid.

14. F. Engels, *Anti-Duhring,* pp. 114-145.

CHAPTER 1

Universe Is Made of Diversified, but Related Matter.

———

OFTEN WE FIND OURSELVES ASKING: Where did we come from? Where are we going to? Did we come into being what we are through one single act, whatever it may have been, or whatever may have caused it? Or are we the unfinished and still changing product of millions of material (physio-chemical biological), step-by-step evolutionary processes? Where do we stand with respect to every material thing in life, i.e., our immediate environment; with respect to other human beings-black, brown, yellow, and white races; with respect to plants, animals, oceans, mountains; with respect to the planet Earth on which we have been Jiving; with respect to the rest of the solar system, and for that matter the rest of the universe? If we were to pay a visit to some of the largest zoos or plant and flower nurseries in the Unites States, what would we be able to see? We could observe that there are numerous varieties of fascinating animals and hundreds of romantically beautiful flowers, all of which are highly detailed, possessing individualized distinctions and characteristics. Every animal, plant, and flower is so unique and beautiful in individuality and character that one would rightly become absorbed in material wonders of their distinctions. How ever, the more important question to ask is: Does every individual flower, animal, living being, as uniquely distinguishable as they are, really Jive in isolation from other plants, flowers, animals, other material things, and the general environment in which we all live?

As highly individualized and diversified as they are, in terms of their looks, functions, and modes of existence, how are they related to one another? Are they necessary relatives of one another, some very distant and some very close? And if they are, do they all *constitute a big, variously related, unified family?* And if they do, what do they have in common? What do they share while maintain doing their individualities and distinctions, that could be used as a basis for establishing any material (life) relationship?

We are told by modem natural scientists, physicists, chemists, biologists, astronomers, and cosmologists that there are billions of other solar systems in addition to our own that the majority of us have been unaware of. Some of them are said to be bigger, some are as big as our own solar system, while others may be smaller. That makes us wonder even more as to what they too are made of; and whether or not we have anything in common with them.

Are there any plants, flowers, animals, and life in general, as we know it, living in these solar systems? And if there are, what would they look like? Would they, in any form or shape, look like us? Or do they look quite different?

Most of our experiences are limited to our own immediate environment, on the planet earth, where we are born and die. For some of us, the existence of our own solar system, let alone the existence of billions of others, appears to be a figment of imagination, simply unthinkable. They are not tangible enough to see, feel, touch, understand and finally recognize. Our actual experiences are limited to a tiny fraction of the universe, i.e., our own planet Earth. Therefore, much of the ideas and opinions we form; even if properly understood, may tend to be narrow-minded in depth and scope. Let us pursue our discussion by attempting to under stand our own life and our own experiences.

How do we materially experience life? Setting momentarily aside all the luxurious conveniences that we enjoy, and concentrating on the basics of life, we notice that human beings in order to survive, need food, nutrition, and shelter, despite their quality. These are so basic and essential for human survival that they are usually taken for granted. Then what is the food. i.e., the vegetables, beef, water we drink, and literally thousands of food items that we nowadays get from supermarkets, made of?

Based upon thousands of years of human experiences, we discovered that we have to have oxygen in order to survive. Life without oxygen is impossible. We have biologically experienced the use of oxygen for millions of years. Only lately, we have be come conscious as to how important it is in guaranteeing our continued survival. Now that we know how important this element is to our lives, we should ask: What are oxygen we breathe, the beautiful beaches in which we swim and relax, the sunshine under which we try to get a copper tan, the food we eat, the school buildings in which we sit and study, the car we drive, gasoline we consume in getting to work, school, recreation places, the pages of this book that you are reading right now, and also the infinite number of other items we encounter, use and misuse in our daily life, yes, all of these and many million others, made of? *They are made of matter.*

One could go on and name every individual material thing we experience as we go through life. The list would be simply infinite. The point being that all of these things are material and objective; we can see, feel and touch upon encountering them. They are being made of some stuff. Do we see, feel, and touch all material things? No, there are some "matter," i.e., minute par ticles that cannot be seen with naked eyes. But there are electronic microscopes that can magnify these atomic particles about two million times and enable us to get much closer to them, and therefore understand them.

To engage in investigation toward knowing these phenomena means to find out what they are exactly materially made of; how they exist, change, develop; and above all how they are related to one another. And, that is exactly what natural sciences such as physics, chemistry, biology, and more specialized sciences derived from these, have been trying to do.

To have something to go by, the natural scientists have been using the term "matter" to stand for all objective and material things. Matter is what everything is made of.

Based upon the same reasoning, philosophers who think that the answers to finding out what everything is made of can only .... be found in the material objects themselves, as opposed to cook ing up something in our minds and attempting to impose upon them, these philosophers, too, have been using the philosophical term "materialism." It simply means they use reasoning that is directly related to these objects. Today, no science could even move one inch forward without employing materially oriented reasoning. For example, if your body is developing a growth, your doctor would conduct a biopsy of that specific part of your body, and after having completed careful a laboratory examination as to the nature of that growth, he would determine whether or not it should be removed. His examinations, among other things, would involve seeing the germs under the microscope. He does not go and consult his Bible or Koran or any other religious book as to how be can determine whether or not that growth is malignant. For that he is expected to have studied biochemistry and biology for years. I'll bet you would not want to be the patient under going surgery if the physician consulted the religious books to find an answer to your medical problem, instead of relying upon his medical knowledge and practice. Now, what is wrong with that?

Please do make a distinction between the religious and scientific understanding of materialist reasoning, or philosophical materialism. The latter is the orientation of this book, to be further explained as we progress in developing our ideas, while the former (religious understanding of materialism) is weII disclosed by the following quotation from F. Engels:

The fact is that Starcke, although perhaps unconsciously, in this, makes an unpardonable concession to the traditional philistine prejudice against the word· materialism resulting from the long-continued defamation by priests. By the word materialism the philistine understands gluttony, drunkenness, lust of the eye, lust of the flesh, arrogance, cupidity, avarice; miserliness, profit-hunting and stock-exchange swindling-in short, all the filthy vices in which he, himself, indulges in private. By the word "idealism," he understands the belief in virtue, universal philanthropy and, in a general way, a "better world" of which he boasts before other, but in which he himself at the utmost believes only so long as he is going through the depression or bankruptcy consequent upon his customary "materialist" excesses. It is then that he sings his favorite song; "What is man? Half beast! Half angel!"[1]

1.-MATERIALITY AND OBJECTIVITY OF MATTER

There is no matter that is non-material or non-objective. Everything is made of some stuff, things which are observable, feelable, touchable, and perceivable. Matter, according to modem sciences, can neither be destroyed nor created by non-material things.

2-THE CONCEPT OF MATTER DISREGARDS INDIVIDUAL DISTINCTIONS.

We already explained that there are an infinite number of individual objects that we experience in our daily lives. Are they all the same thing? Our superficial daily observations would make us conclude that they are not. How can anybody confuse a car for his dog, or his television set for his dinner on the table? Since we recognize distinctions among individual objects, are we justified in absolutifying the individual distinctions to the point where no relationship of any kind can be established between them? To say that all these things are the same would be doing injustice to our daily commonsense experiences. Because we see things that :are both qualitatively and quantitatively different. Yet, to conclude that these differences are of absolute, totally different natures would be tantamount to ignoring what these qualitatively different things have in common, and how they are materially related to one another.

So the general concept of matter, while recognizing ma materiality, objectivity, and commonality of material things, tends to stop short of pointing out the distinguishing features of different kinds of matter, their actual state of being, their connections, interconnections and interrelationships. These important concepts will get more comprehensive treatment as we progress in developing our main ideas.

3.-SOLID, LIQUID, AND GASEOUS ENTITIES AS THE MOST GENERAL FORMS OF MATTER

In how many different forms does matter manifest itself? Well, so far there are three major forms in which matter can be seen, felt, and touched as experienced by mankind. However, in finite individual things fall under one of these three major categories. First, solid matter is made of "hard things," i.e., rocks, pebbles, mountains, and the car we drive. Second, liquid matter is anything \hat is liquid-like, such as water, the shampoo we wash our hair with, the medicine we take when we are sick, and so forth. Third, gaseous matter is less· tangible than the other two. Some common types of this matter are the gas we use for cooking, and oxygen we breathe for our continued survival.

1. F. Engels, *Ludwig Feuerbach and The Outcome of Classical German Philosophy,* pp.31-32.

The Concept of Matter Is Too Broad, Too Block-Like to Represent Material Universe; There Is a Need to Introduce a Smaller Building-Block (Atom).

———

IN ORDER TO UNDERSTAND MATERIAL life, it is necessary but not suf fi- cient to use the concept of matter, even without certain qualifica tions. But the concept of matter is too broad to stand for all individual instances of material life, because matter manifests itself in an infinite number of ever-changing contents and forms. The concept of matter does not tell us anything about what *matter itself is made of and what constitutes its inter- nal structure.* It is too broad to account for an equally important aspect of material life, i.e., individuality and distinction. For example, even though both beautiful flowers, such as carnations and roses, as two forms of mate- rial universe are made of matter, it would be a mistake to consider both of these as being the same, just flowers, just matter. Matter is diversified and appears in infinite distinct forms, living in the same family of matter. Just as I cannot be taken for my father because we are two different individu- als, to say that we are two "totally different persons" would be like saying that we have no material, (biological relationship) of any kind whatso ever. While we do have certain particularities, genetic as well as phenotypic, appearance and even temper, we are still related to one another, in every way you can think of, and one is the organic continuation of the other. A rosebush cannot be confused from the earth in which it is planted. The rosebush and the earth are quite distinct from one another. Yet, a rosebush *can only live as a rosebush* when *it* is properly planted, regularly watered while being exposed to the sun. The rosebush, the earth, the water, and the sunshine are nothing but qualitatively different and distinct forms of matter. The rosebush, in order to survive, at least in a blossoming form, needs the material association of the other forms of matter; the earth, the water, and the beautiful sunshine. This definitely shows that nature is a big, diversified family. Relation ship and individuality are two of the most beautiful characteristics of all materially existing, living things in the universe. That is why nature is an open system of diversified relationships or profoundly interrelated diversities.

1.-Atomic Structure As the Smallest Building-Block; As a Unit of Atomic Universe

The concept of matter points to materiality and objectivity of material universe. But we can not go around declaring that every thing is material and objective without attempting to investigate as to what everything is specifically made of and what, above all, constitutes its internal structure and the more basic component parts of its composition. We need much smaller units of investigation of which the material universe, including the most organized forms of matter, namely, human beings, is made of.

Modem natural sciences tell us that these smaller building blocks are atoms. The concept of atom is not new; it was first introduced by ancient Greek philosopher, Democritus. But, it has never been so important in resolving practical problems of modem societies as it is today. Our era is characterized by gradual under standing and practical use of what atoms offer and how they certainly facilitate life for all of us. I have no doubt that modem production would force us to go atomic all the way through in time. We, therefore, need to develop a mental orientation and preparation in order to readjust our thinking to this wonderful era in order to make an easier transition. The use of atomic wonders is more noticeable in our army defense capabilities than in other areas. The existence of nuclear weapons, that· in more evolved forms, would be capable of blowing up the entire planet earth into pieces, should some madmen in Washington decide to push a few buttons, is a clear example of what the atomic age can do. But, its peace-oriented practical uses in other walks of life could produce wonders never experienced by man before.

This smaller building block is the atom. Atoms are very small, microscopic, and cannot be seen with naked eyes. Atoms are more basic and fundamental, accounting for the physiochemical, and biological compositions of material universe. Are all atoms the same? No, not necessarily. Atoms have different, individualized internal structures ad compositions. Are there any particles that are smaller than atoms, which could be used as more fundamental building blocks of the universe ? Of course, there are, but not perhaps as units. We are told that 95 percent of the universe is made of one given atomic structure, namely, hydrogen. So far 135 different atomic structures have been found. Some of the most common ones are gold, silver, oxygen, hydrogen, and so forth.

The future will testify to discoveries of other presently un known atomic elements (compositions), and the universe, in time, will certainly evolve others. Since the atomic compositions of these 135 known elements cannot be the same and, therefore, con fused with one another, then individualized distinctions of different atomic compositions become important. There is a need for investigating particularities of atoms. And this arises from the fact that there are different kinds of atoms in terms of quality-quantity relationship. But these particularities of atoms are not absolute *(totally unrelated nature)* and should not, therefore, be treated as such.

2.-ATOMS AND MOTION

If we look around, we see that our planet Earth, being a con glomeration of many different kinds of atoms, encompassing all the 135 known elements and still many others not found as yet, as a unit, revolves around the sun, completing its revolution in 365 days, at the speed of 1500 mph, while spinning around its own axis, and completing it in twenty-four hours at the speed of 1,000 mph. The moon also. revolves (moves) around the Earth approxi mately once every 30 days. That goes for other planets of our solar system as well. *They all move around a relatively patterned orbit.* These planets are nothing but conglomerations of literally infinite numbers of atoms in ceaseless motion.

3.-MOTIONS ARE PRIMARILY INHERENT IN ALL ATOMS

Speaking about integrated atoms forming big bodies, are they the only ones that move? What about individual atoms? Do they have any movement of their own, especially individual internal **movement?**

John J. Carlin, expounding on the kinetic-molecular theory of matter, explains molecular movement (motion) in three different states, gaseous, liquid, and solid, as follows:

... in the gaseous state, the molecules of a substance are moving at high speed and have a large amount of space be bettween them. As the temperature is lowered (and/or pressure is increased) the molecules tend to slow down and come closer together, finally forming a liquid. Further cooling of the liquid causes the molecules to slow down still more and permits them to come still closer together until they form the solid state. In the solid state, molecules may be thought of as merely under going slight vibration. It can be shown that molecules · are moving in the solid state by the experiment of placing a lead plate on top of a gold plate. After sometime, molecules of gold plate may be found in the lead plate and the molecules of lead in the gold plate ...1

Consider the following illustration which almost everybody can relate to. If you were to be sitting in a dark room, with heavy curtains covering all windows, assuming it is a beautiful sunny day-then, if you were to create an opening through one of the windows, so that a sun ray would be coming through that particular window, you would see that the contrast between the darkness and the light created by the sun ray would make the small particles in the space in the room exposed and, therefore, observable to the naked eye. These individual particles are in ceaseless motion. Or, on a very dark evening turn the headlights of your car on. The light beam exposes the particles floating around.

What about the wood out of which your dining table is made? Or the metals out of which your car is built? Or the material out of which the desk at which you are studying is constructed of? All of these are examples of solid integrated atoms. They seem to be stationary and non-moving. And if they are left in a given place, not touched by anyone, do they not remain unmoved? Indeed, there is atomic and molecular movement (motion) in every solid or macro-body, even though it may not be observable to the naked eye.

The conclusion we arrive at is that the atomic universe, regardless of its atomic composition, regardless of its individualities and distinctions, has always been and will continue to be in cease less movement and motion. For indeed, there is no such thing as motionless atoms, or atom-less motion. The two go hand in hand.

Do all atoms have the same kinds of motions and movements?

4.-MOTIONS AS GENERAL MODE OF EXISTENCE OF ATOMS

The fact that every atom is accompanied by motion or com combinations of motions shows universality of motion. But, not every atom or conglomeration of atoms, molecules and macro• bodies have the same amount and complexity of motion. So in this sense, the existence of motion on a universal basis is "absolute." By absolute, I mean that there is no atom without motion and no motion without atoms. Motion is inherent in all atoms.

The fact that every atomic structure has its own complexity and level of organization, necessitates a discussion of individuality and relativity of motion. This simply means motions as related to particular and concrete atomic, molecular, and macro-body structures. Albert Einstein, in his theory of relativity vs. "absoluteness" of motion, as related to specific parts of material universe, holds:

Mathematics deals exclusively with the relations of concepts to each other without consideration of their relation to experience. Physics too deals with mathematical concepts; how ever, *these concepts attain physical content only by the clear determination of their relation to the objects of experience.* This in particular is the case for the concepts of motion, space, and time. The theory of relativity is that physical theory which is based on a consistent physical interpretation of these three concepts. The name "theory of relativity" is connected with the fact that motion, from the point of view of possible experience, always appears as *the relative motion of one object with respect to another* (e.g., of a car with respect to the ground, or the earth with respect to the sun and the fixed stars). Motion is never observable as "motion with respect **to space," or as it has been expressed, as "absolute motion."** The "principle of relativity," in its widest sense, is contained in the statement: *The totality of physical phenomena is of such a character that it gives no basis for the introduction of the concept of "absolute motion"; or shorter but less precise:* There is no absolute motion.2

We, therefore, have to introduce the concept of *self-motion* to stand for specific existence and individual internal structure of atoms. We will deal with this most fascinating subject more comprehensively after we have talked about different forms of motions.

1. John J. Carlin, *Learn Chemistry The Easy* Way (Bronxville, New York: Cambridge Book Company Inc. A subsidiary of Cowles Communica tions, Inc. 1960), p. 3.
2. Albert Einstein, *Out of My Later Years* (Secaucus, New Jersey: The Citadel Press, 1974), p. 41.

CHAPTER 3

Forms of Motion

———

OUR. UNDERSTANDING OF THE NATURE of motion, like everything else, has gone through an evolution. The most widely experienced type of motion is when an object big enough to see, feel, and touch strikes against another and, therefore, causes motion or movement. In this type of motion, one object, through the action and interac tion of others, moves from one given place to another, such as the interaction of our planetary system, keeping the planets in move ment through the laws of gravitation. This seemingly is an example of mechanical motion, the cause of which is external.

But then material existence (life) is not reduced to this type of motion only, motion associated with observable big bodies. Objects must be broken down to smaller parts, such as molecules and atoms. One must establish motion (mode of existence) on atomic, molecular, cellular, and organic levels.

According to F. Engels, every type of motion is being specialized by a given science. For example, mechanics studies the motion of macro-bodies, physics deals with motions of molecules, and chemistry investigates motions related to atoms, while biology attempts to understand motions of the most developed, complex parts of existence, namely biological organisms from the simplest cell to the biology of man and his brain. This even solves the mystery of what is the nature of our natural sciences, i.e., physics, chemistry, biology, and other more specialized sciences derived from these.

Even though it is convenient for man to say "this type of motion is mechanical, while those are physical, chemical, and finally biological," and "benefit" from these arbitrary distinctions in solving daily problems, in real life, these different types of motions are not separated from one another, as we mistakenly think they are. They are organically integrated and certainly transform into one another in the process of evolution. The fact that there is a diversity of distinct types of motions *should not give us the right to deny their mutual dependency and interrelationships.* The fact is that in objects one or several types of motion may be predominant, while others may play a subordinate role. In short, it is the combinations of types of motions which differ in different parts of the material universe. These combinations are not fixed; they are in a state of flux.

This shows our earlier claim that our self-motioned universe constitutes an infinite number of ever-changing diversities and that evolutionary processes develop their own particularities of evolution, while always dynamically entering newer relationships.

1.-NATURE ITSELF IS CONSTANTLY OVERLAPPING. THEREFORE, SEPARATE TREATMENT OF MOTIONS IS UNSCIENTIFIC AND BAD PHILOSOPHY.

The futility and inadequacy of separate treatment of motions is today testified to by an ever-growing tendency in the studies of natural sciences to be forced to cross and overlap one another's traditional boundaries. This is a result of the realization of the inseparability of different kinds of motions, their distinctions, active. interaction and interrelationship. *Nature itself overlaps; and so should the. natural sciences reflecting it.* This is demonstrated by certain hybridization of different natural sciences in order to ex plain more complicated phenomena, such as for example, physio chemical-biological aspects of our troubled, over-polluted environment.

Specialization of one or several kinds of motions is not enough. We are required to pay equal attention to their complex interrelationship, the role they play individually and collectively in the process of individual and universal evolution. Now we attempt to introduce what I think the most important concept in the entire book, whose understanding is very essential in order to understand the universal processes of evolution, namely, the concept of self-motion.

2- SELF-MOTION AS THE BASIS OF SPECIFIC AND INTERNAL FORMS OF EXISTENCE OF ATOMS (SELF-MOTIONED ATOMISM)

The first question to ask is: Where is the motion associated with every individual atom coming from and what is it caused by? At this time, in order to minimize confusion, we would keep the explanatory examples on the atomic level.

Usually when we think of motion, we think of, say, one atom striking against another, and therefore causing motion and move ment. This can be classified as *external,* because one atom, coming from someplace else, hits another one. The assumption is that the first atom would not have moved unless it was struck by an other which, basically, caused the motion and the movement. There is nothing wrong with this kind of reasoning, because we see in our daily life a lot of things hitting against other things and, therefore, causing motion and movement.

A classical example of an *externally produced motion* is the well-known example of billiard balls. Whoever has played billiards can vouch for the fact that no ball would move (be in motion) on the table unless it was struck by another ball.. Let us follow the number of observable motions produced in the process of one shot. There are at least three steps to produce a given mechanical motion. By mechanical, I mean one object hitting another, and thereby, causing a motion. First, the cue ball is aimed at and struck by the player. Second, the cue ball, being struck, transfers the motion to the eight ball. Third, the eight ball, being hit by the cue ball, moves (is put into motion) into the billiard pocket.· These three simple steps, combined, show the process of one mechanical motion, which is quite observable and easily reproducible, all created and caused on an external basis. In this example, every entity received its motion from another, or else they would not have moved.

What happens when players quit playing and the balls come *to* a standstill position? If the balls are left alone on the table, *would they not remain in eternally motionless position?* How can we claim there is any motion at all? We already claimed that motion is inherent in matter. And yet, here is a clear example where according to our sense of sight, "the balls do not move at all."

Doesn't that knock out our claim that atoms and conglomerations of atoms, and motion are inseparable? Of course not! This only shows the inability of our sense of sight to see and observe finer, more subtle, and more fundamental forms of motions, *namely, those of atomic and molecular ones.* Let us go back to our earlier line of question: What is everything made of? In this case, what are the billiard balls, the cue ball, the table, the floor on which the table is located-made of? *They are made of literally billions of individual atoms, every one of which has its own individual,* internal**, ceaseless atomic movement and motions.**

John Tyler Bonner, in his book, *The ideas of Biology,* brilliantly reveals the nature of the material universe in the following terms:

The molecules, from the smallest, such as hydrogen, to the largest proteins, are made up of protons, electrons, and neutrons. These particles, with all the other inside the nucleus, are ubiquitous. This means that living organisms are not alone in being made up of electrons, protons, and neutrons. This is true of all matter. A table, a glass of water, a layer of dust, or the slate of tombstone are nothing more than masses of these three particles.1

3.-ATOMIC MOTION IS NOT PRIMARILY EXTERNALLY CAUSED. IT IS SELF-CONTAINED (SELF-MOTIONED ATOMS).

Albert Einstein, in his book *The Evolution of Physics*, employing a quotation from Brown, *clarifies the source of motion*. He (Brown) reports:

That while examining the form of these particles immersed in water, I observed many of them evidently in motion ... The motions were such to satisfy me after frequently repeated observations that *they arose neither from current in the fluid nor from its gradual evaporation, but belonged to the particle itself.'*

Einstein himself, talking about "what wave is," substantiates the inseparability of atoms and motion and, therefore, adds:.

The observed motion of the wave is that of a state of matter and not matter itself ... *The essentially new thing here is that for the first time we consider the motion of something which is not matter, but energy propagated through matter.'*

John J. Carlin, elaborating on the kinetic-molecular theory of matter, maintains:

According to the kinetic-molecular theory of matter, a pure substance is composed of small particles, called molecules (molecules are combinations of individual atoms) which have the properties of the mass (recognizable quantity of the material). *The molecules of the substance are in continuous, never-ending motion."*

And, lastly, F. Engels crowns the subject by saying that:

The whole nature accessible to us forms a system, an inter connected totality of bodies, *and by bodies we understand here* all material *existence extending from stars to atoms,* indeed, right to ether particles, in so far as one grants the existence *of the last named. In the fact that these bodies are inter connected* is already included that *they react on one another, and it is precisely this mutual reaction that constitutes motion.* It already becomes evident here *that matter is unthinkable without motion."*

This was stated one hundred years ago, showing that F. Engels was way ahead of his time.

At this time, we get more specific as to what every atomic structure is composed of. Every individual atom is made of at least three major parts; protons, neutrons, and electrons. There are other sub-atomic particles which are subordinated to these three major component parts of atomic structure and their interrela-. tionship.

1. John Tyler Bonner, The *Ideas of Biology* (New York: Harper Torch.. book, Harper and Row Publishers, 1969), p. 3.
2. Albert Einstein. *The Evolution of Physics* (New York: A Touchstone Book, Published by Simon and Schuster, 1938), p. 59.
3. Ibid., pp. 100-101.
4. John J. Carlin, *Chemistry The Easy Way,* p. 2. 1970.
S. Frederick Engels, *Dialectics of Nature* (New York: International Pub-lishers, 1967), p. 36.

Some General Observations and Conclusions on the Structure and Behavior of Self-Motioned Atoms

---

THE COMPONENT PARTS OF ATOMS, such as protons, neutrons, and electrons, while going through a never-ending, but always chang ing combinations of motions while interacting, constitute a unit, i.e., an atom, or simply a unified atom.

The component parts of atoms possess materially opposing tendencies. These opposing tendencies manifest themselves through all individual atoms being, to various degrees, electrified and magnetized. The protons, for example, are positively charged, while the electrons are negatively charged. A neutron may be considered as being composed of equal number of protons and electrons. It, therefore, has the characteristics of both. So we notice that electricity and magnetism are inherent properties of every individual atom, making up its mode of existence.

According to F. Engels:

... In the fact that these bodies are interconnected is already included that they react on one another, *and it is precisely this mutual reaction that constitutes motion.1*

So then, it is the interaction of these opposing particles con tained in every individual atom that constitutes motion. These motions are not primarily caused or generated by outside atoms.

They are self-contained and inherent in all individual atoms, re presenting their mode of existence. Individual atoms are, there fore, primarily self-motioned and self-caused, and need not be moved from outside by other atoms.

Self-motioned atoms do not have fixed, unchangeable sealedoff boundaries. The internal composition of self-motioned atoms does not remain unchanged eternally. They are in constant trans formations.

1.-INTERRELATIONSHIP OF SELF-MOTIONED ATOMS

The fact that we talked about individualized self-motioned atoms does not mean that they can have absolute, complete, individualized existence, living by themselves in isolation from the rest of the self-motioned atomic environment.

Self-motioned atoms are mutually penetrative, and mutually dependent for their continued and ever-changing existence and development.

2.-SELF-MOTIONED ATOMS HAVE VARYING COMBINING ABILITIES, FORMING MORE COMPLEX SELF-MOTIONED ENTITIES.

It is the interpenetrating and combining abilities of self motioned atoms that would result in the evolution of more com complex material entities.

Several atoms combine to form a molecule and many molecules combine to form a more complex body, a cell. Billions of cells combine to form a still more complicated form of life, i.e., a plant, animal, and human being. We talked about individuality of self-motioned atoms in order to pinpoint the *primary source of motion.* To that extent, we are interested in the individual posture of atoms, but since there is nothing in the entire universe that is made up of one single atom, and everything is of multi-self motioned atomic character, therefore, we concern ourselves with collective existence and behavior of self-motioned atoms.

In a sense, modern physics is more interested in collective atomic behavior than in their individual existence. On this there is an identity of views between F. Engels and Albert Einstein.

Einstein maintains that:

Quantum physics formulates laws governing crowds and not individuals. Not properties, but probabilities described, not laws disclosing the future of systems are formulated, but laws governing the changes in time of the probabilities and relating to great congregations of individuals. •

In the following statements, F. Engels makes a distinction between self-motioned atoms, molecules, and complex molecules having their relatively individualized modes of existence with given qualitative compositions, as smaller units. At the same time, their combination and interrelationship results in more complex forms of matter with given quantity-quality relationships, very much different from the individual parts, namely, self-motioned atoms, molecules, and complex molecules:

The molecule is decomposed into its separate atoms, which have quite different properties from those of the molecule. In the case of molecules composed of various chemical elements, atoms or molecules of these elements themselves make their appearance in the place of compound molecules; in the case of molecules of elements, the free atoms appear which exert quite distinct qualitative effects: the free atoms of nascent oxygen are easily able to effect what the atoms of atmospheric oxygen, bound together in the molecule, can never achieve.3

But the molecule is also qualitatively different from the mass to which it belongs. It can carry out movements independently of this mass and while the latter remains apparently at rest, e.g., heat oscillations; by means of change of position and connection with neighboring molecules it can change the body into allotrope or different state of aggregation.•

A very common example of this is the composition of water. It is made of two volumes of hydrogen and one volume of oxygen. These elements have their own individualized self-motioned atomic structures, qualitatively different from one another. Yet, they combine to form a still qualitatively different thing, namely, water. This indeed is a beautiful example of the individuality and relation ship of just one part of the material universe.

Relating to the general subject, Engels goes on to add that:

It is just the same with cause and effect; there are conceptions which only have validity in their application to a particular case as such, but when we consider the particular case in its general connection with the world as a whole *they merge and* **dissolve in the conception of universal action and interaction.** *5*

3.-SELF-MOTIONED ATOM IS DIALECTICAL IN NATURE.

Every self-motioned atom, while having its own individual structure, its own inherently internal motions as a unit, is materially interrelated to other self-motioned atoms..Every self motioned atom provides conditions for other self-motioned atoms to exist and develop, while reciprocally receiving material conditions for its own development and evolution. Therefore, we have to talk about self-motioned inter-atomic connections, penetrations, and relationships.

Interrelationships can not be cooked up in people's minds.

They must be actively sought and discovered from the actual ma-material being of the self-motioned atomic universe.

To establish all the ever-changing connections, inter-interconnections, and interrelationships of the evolutionary processes of the material universe on sub-atomic, self-motioned atomic, molecular, and macro-body levels means to establish their dialectical nature.

4.-Evolution of Self-Motioned Atoms

Self-motioned atoms are not internally fixed and unchangeable with the same repetitious circular motions. Self-motioned atoms as well as self-motioned inter-atomic and intermolecular bodies are ceaselessly experiencing change. In short, the entire self motioned atomic universe is going through transformations and evolutionary processes.

If we were to trace the origin of our solar system, we would see that there is material evidence presented by modem natural scientists testifying to the fact that approximately four billion years ago our solar system came into being. At that time, the sun is said to have been much hotter than it is today. The other planets, including our planet, Earth, were in molten states. All the vegetation, trees, all kinds of animals, including human beings were non existent. It took a couple of billion years for the sun's temperature to drop, making the other planets cooler, when the formation of atmospheric oxygen became possible. There and then, must have come about the very primitive forms (germs of life) of vegetations. The process of evolution made the emergence of more complex forms of life possible. The evolution of the cell, a self-regulated organism, was an important step in the process of evolution. John Tyler Bonner comments on the makeup of material **universe:**

Even though a cell is small, it contains a very large number of molecules, for molecules are themselves so small ... In fact an average size cell contains 200 million-million (2.10^{14} molecules ... Molecules, from the smallest, such as hydro gen, to the largest proteins are made up of electrons, protons, and neutrons."

This closes the gap between what some of us mistakenly call "living vs. non-living matter" in which we consider them as being two "absolutely different things," as if there were no material relationships between the two. They are both life in different forms, with different levels of evolutionary complexity, and physio chemical-biological organizations. The combination of certain self-motioned atoms resulted in the evolution of molecules and the latter, through complex evolutionary processes, evolved into a cell. The one-cell organisms evolved into many-cell organisms and from there on the process of diversifications and biological complexities was on the road.

P. Anderson profoundly testifies to this intimate material commonality and interrelationship of the so-called "living and non-living":

... but no serious thinker now believes that life is a separate kingdom within the universe ... The same laws that prevail in the atom and molecule must command the cell, the animal, and the brain.[7]

D. S. Halacy, Jr., a noted American author on the science of genetics, illuminates, in general terms, the evolution of our solar system and life in general:

Until about a century ago, it was commonly thought that the earth was only a few thousand years old. This belief stemmed from a strict chronological interpretation of the Bible; Bishop Usher carefully added ages all the way back to Adam and established the date of creation as 4004 B.c. In 1785, Scottish naturalist, James Hutton published his *Theory of the Earth,* in which he proposed that natural forces such as weathering, erosion, river channeling, and the like had been proceeding at a uniform rate for ages. This concept appealed to Britisher, Charles Lyell, who wrote the important book, *Principles of Geology*. This coming of geological science, and the development of methods for estimating the age of the earth, made it evident that earth must be much older than 6,000 years.... Buffon suggested that the earth was formed when the sun collided with a comet. It was soon learned that such a collision would not be very violent, and this theory was dropped. However, Isaac Newton later suggested that the entire solar system began when a cloud of cosmic gas and dust collected, condensed, and collapsed to form the sun and its satellites. This is still accepted as pretty much what happened, although Newton's picture was not complete or correct in every detail. Many brilliant men, including Immanuel Kant, Pierre Simon de Laplace, Herman Helmholtz, Lord Kelvin, James Clerk Maxwell, Sir James Jeans added their suggestion. . .. Charles Darwin greeted the suspected antiquity of the earth with relief and apprecia **tion, for it gave more time "for natural selection" to have** worked. Since then, of course, the earth bas been recalculated to be older and older until at present it is generally thought to be about 4.6 billion years old. Lunar materials seem to be about the same age and so strengthen the estimate for earth. . . . Life is not as old as the formation of the hot cloud of gas that would become our planet, unless it was something quite different from what we are familiar with. It is doubtful that even the weirdest kind of evolution could account for creatures existing at a temperature of thousands of degrees. The coming of life must have had to wait for at least some minimum period and cannot span the full 4.6 billion years. . . . According to Oparin, a Soviet scientist, amino acids and protein formed spontaneously. Then these molecules slowly aggregated into "colloid" or gelatinous grouping. After hundreds of millions of years, such aggregations with the property of cells developed."

t. F. Engels, Dialectics *of Nature* (New York: international Publishers, 1967), p. 36.

2. Albert Einstein, *The Evolution of Physics From Early Concepts to Relativity of Quanta* (New York: Simon and Schuster Publisher, 1938), p. 297.

3. F. Engels, *Dialectics of Nature*, p. 28.

4. Ibid., p. 29.

S. F. Engels, *Anti Duhring,* pp, 23-33.

6. John Tyler Bonner, *The Ideas of Biology,* pp. 2, 3.

7. Paul Anderson, *Is There Life On Other Worlds?* (New York: The Crowell· Collier Press), p. 42.

8. D. S. Halacy, Jr., *Genetic Revolution* (New York: A Mentor Book, New American Library, 1974), pp. 38-41.

CHAPTER 5

Self-Motioned Atomic Universe
Is Law-Governed.

———

GEORGE GAYLORD SIMPSON, QUOTING A passage from the argument of the modem molecular biologists in claiming the applicability of physio-chemical laws to biology, states that:

For biology, generally, probably the most important single conclusion reached to date is the conviction that living systems do *indeed obey the physical and chemical laws that govern the rest of the Universe,* that the detailed working of the living organism is amenable to exploration by physical and chemical probing, and that the *properties of living organisms are totally comprehensible in chemical terms.'*

By the way, there is nothing mysterious about the so-called "physico-chemical laws." By that is meant laws that are applicable to "non-living matter," the earth, mountains, water, just to mention a few, as opposed to "living mater," biological organisms.

The point at this time is not to prove that physico-chemical laws are "totally' applicable to biology with or without certain modifications, but rather, whether or not the self-motioned atomic universe is law-governed, as opposed to being completely chaotic.

Lincoln Barnett, concerning the law-governed nature of the universe, quotes Dr. Albert Einstein as follows: "Einstein more than once expressed the hope that the statistical method of quantum physics would prove temporary expedient. 'I cannot believe,' wrote Einstein, "that God plays dice with the World" ... He believed in a universe of order and harmony.' "[2] .

The preceding statements, plus others in our previous discussions, point out that our self-motioned atomic universe, regardless of its multiplicity of forms of existence, moves, evolves, and develops in accordance with certain laws and not chaos. Let us first of all discuss what we mean by laws of the self-motioned **universe.**

1.-LAWS AS INVOLVED, COMPLICATED INTERACTION AND RELATIONSHIPS OF SELF-MOTIONED ATOMIC, MOLECULAR, AND MACRO-BODY ENTITIES

Let us build up a general understanding of what we mean by laws, point out some of their most widely accepted characteristics and then engage in a deeper and more comprehensive discussion of them. On a *common-sense level,* by something being law governed, we mean some relationship that is stable, repetitive, and following a definite course of events. For example, every night we go to bed, we would expect the sun to rise the following day. For that to happen, the sun must still shine, while the Earth, revolving around the sun, must continue to spin, so that the following day, the other half of the Earth would face the sun. The sun has risen for millions of years so far. This shows a pattern of regularity.

Take another example: The conception, birth, development, and, finally, death of a human. being. You first have to have sexual contact between two members of opposite sex, at which time conception takes place. After approximately nine months primarily, genetically determined, complex union and interaction of the male sperm and female egg, the baby is born. It grows, develops, reaches adulthood, and then begins to degenerate, and eventually dies. What do we see in both of these examples? We see that they both show a degree of stability, repetition, pattern, and regularity. All living organisms come into being as they are, pass away and die according to a given pattern, more or less a definite course of events, and tendencies. Since an all self-motioned atomic universe moves according to certain ways, then the universe as a whole is law-governed on a universal basis. Examples could be provided for any part of the universe behaving in a Jaw-governed manner. So far we have shown that stability, repetition, pattern, general tendencies, and universality as some of the common characteristics of Jaws.

2.-·PARTICULARITY OF UNFOLDMENT OF LAW

But in the example of human conception, we notice that a sexual con-tact involving literally millions of genes is needed be fore conception can take place; and in the case of the sunrise, the sun and the Earth have to interact on self-motioned atomic, mo lecular, and macro-body levels, and simultaneously move in certain ways. Then we could safely arrive at the general understanding that *laws have to do with very in-volved interaction and complicated connections and relationships.* A sexual relationship, resulting in successful conception, assuming that the participants are biologi cally normal, can take place between ages up to at least the late forties for women and "?" for men. But no two sexual experiences, even if between the same individuals, are the same. Even though, anytime there is sexual contact, and the result may be a baby, no two babies are, in fact, the same. The age in which the sexuality takes place, the kind of food the mother eats during pregnancy, the kind of physical activity she has,- the physical environment in which she lives, the kind of emotional composition she maintains, whether or not she smokes or drinks, all of these variables and literally hundreds of others influence the unfolding of the seem ingly repetitious, stable, pattern-like, tendency-oriented law, the law of conception. Then we see that laws not only characterize re lationships, but certain kinds of relationships, the ones that *are necessary* in order for *a law to unfold in a particular way.* Since, every self-motioned atom, molecule, and macro-body moves, evolves, and develops, then what really remains stable, repetitious, regularity-oriented, and pattern-like *is their ever-evolving character, and not moments of their existence.* For example, the Earth has revolved around the sun for millions of years, but no two years have been the same. Every revolution of the Earth is quite unique, relatively different from the preceding one, because both the Earth and the sun and, for that matter, the entire solar system are evolv ing. *Thus, the movement takes place in a modified way.*

3.-EVOLUTION OF SELF-MOTIONED ATOMS AND CREATIVE UNFOLDING OF LAW

Unfolding of laws depends actively upon the ever-changing, ever-evolving, self-motioned atoms and their relationship. So laws are dynamic, because the pattern, the regularity, the tendencies, and repetition are creative, *repeating not exactly in the same way, but in a modified, enriched and more involved complex way.* The more involved and developed a given complex self-motioned entity becomes, the more complex and involved the unfoldment of the laws. Therefore, unfolding of laws is relative and transitory also.

4 -OPERATION OF LAWS DEPENDS ON THE COMPLEXITY OF DIFFERENT PARTS OF SELF-MOTIONED ATOMIC UNIVERSE

A cycle in which every definite mode of existence of matter, whether it be sun or. nebular vapor, single animal or genus of animals, chemical combination or dissociation, is equally transient, and *where nothing is eternal but eternally changing, eternally moving matter, and the laws according to which it moves* and changes •.. We are concerned here in the first place with *non-living bodies; the same law holds for living bodies, but it operates under very complex conditions .. '*

Engels goes on:

Our whole official physics, chemistry, and biology is ex clusively geo-centric, calculated only for the earth. We are still quite ignorant of the con-ditions of electric and magnetic stress on the sun, fixed stars and nebulae, even on the planets of a different density from ours. On the sun, owing to the high temperature, the laws of chemical combination of the elements are suspended or only momentarily operative at the limits of the solar at-mosphere, the compounds becoming dissociated again on approaching the sun. The chemistry of **the sun, however, is in process of arising, *and is necessarily* *quite different from that of the earth, not over-throwing the lat-ter but standing outside it.* In the nebulae perhaps there do not exist even those of the 65 elements which are possibly themselves of compound nature. Hence, *if we wish to speak of the general laws of nature that are uniformly applicable to all bodies-from the nebula to man-we are left only with gravity and perhaps the most general form of the theory of transformation of energy,* Vulgo, the mechanical theory of heat. But, on its general logical applica-tion to all phenom ena of nature this theory itself becomes converted into an ***historical presentation of successive changes occurring in a*** *system of the universe from its origin to its passing away, hence, into a history in which at each stage different laws, i.e., different phenomenal forms of the same universal motion predominate, and so nothing remains as continually and uni versally valid except-motion.'*

Here Engels recognizes that self-motioned atomic universe is law-gov-erned,. *but the laws operate (unfold) differently based upon the complexity and levels of evolution of different parts of self motioned atomic existence.* In other words, the unfolding of laws in human biology are more complex than the unfolding of laws, say, in vegetative complex self-motioned atomic and molecular existence; and that laws unfolding in the latter are operating more complexly than less evolved forms of atomic structures. This type of reasoning *is not reductive in nature,* because it recognizes dif ferent levels of self-motioned atomic existence and their corre sponding levels of evolution and development.

5.-LAWS ARE NOT IMMUTABLE; THEY COME INTO BEING CREATIVELY DURING THE PROCESS OF EVOLUTION.

Laws come into being and, therefore, creative unfoldment during the evolutionary processes of self-motioned atomic universe. Unfoldment of laws vary creatively in accordance with level of development of self-motioned atomic structure. *There fore, laws unfold differently in different parts of the universe.* There is no such thing as immutable law, being absolutely and eternally the same, existing separate, and independent from an evolving self-motioned material universe. F. Engels, revealing the very nature of law, comments *on the coming into being and operation of general laws:*

... in one point, however, the history of the development of society proves to be essentially different from that of nature. In nature-in so far as we ignore man's reactions upon *nature-there are only blind unconscious agencies act ing upon one another and out of whose interplay general law comes into operation.* Nothing of all that happens whether in the innumerable apparent accidents observable upon the surface of things, or in the ultimate results which confirm the regularity, underlying these accidents-is attained as a consciously desired aim. In the history of society, on the other hand, the actors, all endowed with consciousness, are men acting with deliberation or passion, working towards definite goals; nothing happens without conscious purpose, without an intended aim. But this distinction, as important as it is for historical investigation, particularly of single epochs or events, cannot alter the fact that the course of history is governed by inner general laws.'

Here Engels brilliantly recognizes that both the self-motioned atomic universe, and as far as the material movement of human society and its corresponding spiritual life of it is concerned, de velop according to general laws. But, more importantly, *he points out that laws do not exist by themselves and that they come into being and operation while the agencies, or what I call self-motioned atoms and molecules interact.* Whereas in human society, conscious actors, representing material and therefore spiritual interests of social classes by deliberation, help bring about a set of material relationships, essentially the way we go about responding to our material needs and requirements (economics) out of *whose interplay material laws of societal evolution emerge.* In the process of evolution, different parts of the universe develop particularities of relationships, and, therefore, particularities of evolution and un foldment of laws. *In short, the actual self-motioned atomic uni verse determines the nature of laws, and the latter actively in fluences the evolutionary processes.* Since nature is creatively and, therefore, variably *overlapping,* with individual parts *flowing into one another,* then laws operative in parts of self-motioned atomic universe, say biological organisms, are materially mediated and influenced by laws operative in other parts of self-motioned nature and vice versa.

This creates the need to talk about the specificities and inter relationships of unfolding of laws.

1. George Gaylord, *Biology and Man* (New York: Harcourt, Brace, Jovanovich, Inc., 1969), p. 6.
2. Lincoln Barnett, *The Universe and Dr. Einstein* (New York; Bantam Science and Mathematics), p. 36.
3 Frederick Engels, *Dialectics of Nature* (New York: International Pub lishers, 1970), pp. 24-28.
4. F. Engels, *Dialectics of Nature,* p. 242.
5. Engels, *Feuerbach and Outcome of German Philosophy,* p. 46.

CHAPTER 6

The Concept of Time Tied Up to General Motion and Movement

———

To UNDERSTAND THE NATURE OF time, let us see what we mean by and how we arrive at time, in a very ordinary common sense way.

If we ask someone what time it is, he would respond by saying, "Two o'clock." What does he mean by "two o'clock"? Why not three or even four-thirty? How do we determine time, and on what principles are our watches based upon? It is by now common knowledge that our planet Earth spins around itself. When half of our planet Earth faces the sun, it is "day-time," and the other side which is not facing the sun makes up the "night-time." A com plete revolution of the planet Earth around itself has been arbi trari-ly divided into twenty-four equal units, or, to put it differently, twenty-four hours. Every unit of hour is divided into smaller units, called minutes, and every minute is made up of sixty seconds. The mechanisms of all watches are based upon the same principle. The *important point to remember at this juncture is that the concept of time is very closely tied-up to the movement and motion of our solar system.*

As far as determining day and night, month and year times, our prin-ciples of determination are simply based upon the move ments of planets, Earth, moon, and the sun. As the Earth com pletes a revolution, around itself, it continues to complete a rev olution around the sun also. By the time it has completed a revolution around the sun, the Earth has completed 365 revolutions around itself.

This shows how .we arrive at the concept of a year time. Again we see that the concept of a year is tied-up to the motion and movement of the Earth around the sun. Based upon the same principle we arrive at the. concept of a· month. The moon also revolves around the Earth. Every revolution of the moon is equal to. l/12 of a "year time." Once again we see that the concept of a "month time" is also determined by the motion and movement of planet, moon around the. planet, Earth.

This, even though common-sensical, is very important since time *manifests itself and is determined by motion and movement.*

1-TIME IS NOT SIMPLY REDUCED TO SIMPLE MECHANICAL MOTION AND MOVEMENT; BUT IT IS INTIMATELY ASSOCIATED WITH EVOLUTION OF SELF-MOTIONED ATOMIC UNIVERSE.

Since we argued before that there is no such thing as motion without matter and vice versa, and that the material universe cannot be reduced. to a simple movement of bodies from one place to another, and that evolution is the inherent characteristic of all material universe, then time, *living through a self-motioned atomic universe,* is necessarily associated with it. *Time is creatively born out of the evolutionary atomic universe,* showing the processes of evolution, the different paces at which the *inter-flowing self motioned universe evolves;* It is quite obvious that different parts of self-motioned, ·inter-flowing atomic universe, while going through evolutionary processes, creatively develop their own level of development, and the paces at which they develop.

2.-PARTICULARITY AND GENERALITY OF TIME

The following statement by the Soviet scientist, V. Rydnik, on the specificity and generality of time is certainly timely.:.

... We have already mentioned the fact that there are two kinds of time in principle: the "proper time" of a body de termined by the physical (chemical) processes in that body, and the "general time" determined by large assemblies of bodies. As a result, just as there is *no space divorced of bodies*, so there is *no time divorced of events.* [1]

Since there is no complete uniformity of the pace of evolu tion relating to different parts of the universe, then time, being associated with evolutionary processes, and actually being born out of it, varies. *Therefore, there are different times relating to different phenomena and events.* Self-motioned, inter-flowing, atomic universe is universally interrelated; therefore, there comes into being a dynamically ever-changing universality of time out of the evolving, and inter-flowing parts and their interrelationships. While the individual parts of the universe and their corresponding times have·a dynamically mutually determining posture, maintain ing their relative independence and level of developments and, therefore, time, they constitute what may be called the generality of time.

Take a romantic love song that may be your favorite. It is made of individual notes, every one of which bas its own individ uality, uniqueness, duration, vibration, and noise quality. No individual note, regardless of its individual expression, can deter mine or constitute the generality (the whole) of the entire song. It is the live and inter-flowing interactions of the individual ex pressions and the embodiment of all notes, going through creative and dynamic variations and progressions that constitute · *what is in process, born and* dynamically evolves as the general quality of the entire song.

3.-Time Is Born Out of, Lives Through, and Is Determined by Self-Motioned, Inter-Flowing Atomic Universe.

Evolving and inter-flowing of particularities of musical notes *gives birth* to the generality of the song. It is no different from particularities of individual times giving birth to generality of time.

Take the following example: Think about two persons that you have known in the last ten years. One who is energetic, aggressive, enterprising, working hard, always doing something, may be going to college, constantly attempting to improve herself ..*r*···and assuming responsibilities. The other being passive, contented as to where she is, never attempting to see beyond her nose, lack- ing initiative, busying herself with a few trivial things. According to our initial measurement of time, in ten years period the Earth has revolved around the sun ten times. If we compare these two different individuals against this measurement of time, we see that what *distinguishes these two persons is the degree of motions and movements that they go through.* The person who goes through more activities (movement) is bound to develop, in any way you think, more than the one with turtle-like movement, who seems to be repeatedly stepping in the same spots. Very little movement, very little development. If we still use the same principle for measuring time, then we see that in the same period of time, the individuals have gone through two different levels of development and, therefore, represent two different times. This really shows that there is no such thing as time being independent of and separate from self-motioned atomic existence, in this case, two highly evolved biological organisms-the two persons.

Therefore, *time is born out of, lives through and is determined by an inter-flowing self-motioned atomic universe.*

Since the process of evolution can only move forward extending into the future, then time, a by-product of the evolutionary process, can only extend into the future and be one-dimensional. There is no such thing as reversing process of development; and there is no such thing as time reversing back into the past.

4.-SPACE IS NOT AN "EMPTY CONTAINER" SEPARATE FROM ITS MATERIAL CONTENT, THE SELF-MOTIONED ATOMIC UNIVERSE.

We fall for this fallacious reasoning that space is an "empty container" because what we see we identify with matter, substance; and what we do not see is "empty space," or "nothingness." But when we include self-motioned atoms, their further subdivision and sub-atomic existence into the concept of matter and substance, then we arrive at the understanding that *there is really no such thing as "empty space."* It is all self-motioned matter, different in quality and form, where bigger bodies, which are conglomerations of individual self-motioned atoms, move around more observably, very much like fish moving around in the oceans. We could not argue that the fish move in the "vacuums," where there is no matter, no water. One form of more evolved matter, fish, move *penetratingly and inter-flowingly* into another one, namely, water. Self-motioned atoms and their more complex forms of existence have length, width, and height which represent three dimensions.

5.--PARTICULARITY AND GENERALITY OF SPACE

But since all parts of the self-motioned universe are in con stant motion and development, then; movement and evolution deprives them of having a well-formed and fixed three-dimen sionality, providing the fourth dimension (motion) which keeps everything in a state of flux, preventing the self-motioned universe from ever attaining a fixed permanent posture and character. Since, self-motioned atomic existence manifests itself in a multiplic ity of forms, then different parts of universe do, indeed, have different forms of spaces.

.Time and space are two. manifestations of life of self-motioned atomic existence and have "no beginning or end" separate from it. V. Afanasyev, commenting on the multi-formity of space, maintains that:

The Russian mathematician, .Nikolai Lobachevsky (1792·

1856) elaborated a new, non-Uclidean geometry, which re futed the metaphysical views on space and extended man's ideas of the spatial properties of bodies. Lobachevsky arrived at the conclusion that the *properties of space are not identical* in different regions of the Universe and that they depend upon the nature of physical bodies and the material processes taking place in them?

1. v. Rydnik, *ABC's of Quantum Mechanics* (Moscow: Mir Publisher, !968), p.345.
2. V. Afanasyev *Marxist Philosophy* (Moscow: Progress Publisher, 1968), p. 67.

CHAPTER 7

Laws of Evolutionary Development of Self-Motioned Atomic Universe

---

IN OUR PREVIOUS DISCUSSIONS, WE talked about the universal evolution of the self-motioned atomic universe. However, it is not enough to proclaim that the universe, in parts and as well as "the whole," is constantly changing, evolving, and developing. It becomes necessary to establish the laws according to which the processes of evolution and development take place. F. Engels claims to have been the first to formulate what is called the general laws of universal evolution or development, or what I alternatively call, universal laws of self -motioned nature or universe.

They are: (1.) The law of unified, interpenetration of op posites, (2.) the law of qualitative transformation, (3.) the law of negation of negation.

I have made slight modifications in the terminology employed for the laws of dialectical materialism as originally formulated by F. Engels in order to accommodate for the concept of "self-motioned atomism," or self-motioned atomic dialectical materialism, which I have introduced and elaborated upon. To that extent, I assume full responsibility for any scientific value or lack of it. As we go along, we will provide a more detailed explanation of what these laws represent and how they apply to different parts of self-motioned atomic nature.

I.-The Law of Unified and Opposing Interpenetration of Self-Motioned Atoms

Once again we start our analysis of this law by attempting to understand the nature of material existence by analyzing what we have been calling the smallest building-blocks of the universe, namely, atoms and their internal compositions. Even though there are more than one hundred sub-atomic particles found, neverthe less, we speak of an atom as more or less a "unified" structure. In the composition of every individual self-motioned atom, we are told that there are three major parts; protons, neutrons, and elec trons, including, of course, the nucleus of atomic structure. They are distinguished by the fact that they have opposing tendencies. For example, protons are said to be positively charged, neutrons have an equal number of positive and negative charges, while elec trons are negatively charged. The south and north poles of magnetized self-motioned atoms also testify to the existence of another phenomenon, namely, the opposing forces of magnetism. There is no moral value attached to these opposing tendencies. There is also nothing special about the terms "positively charged," "negatively charged," or "north" and "south" poles of magnetized atoms, other than the fact that they represent certain opposing tendencies and forces. Positive and negative charges are dual opposing characteristics of the phenomenon of electricity. These opposing tendencies are found in atoms, aggregations of atoms, molecules, and more evolved bodies.

But the nature of opposing tendencies changes as a given self-motioned entity evolves, becoming more complex in structure. Opposing parts of atomic structure are not disjointed, existing separate from one another. They constitute a closely related "unit." In other words, there is a unified opposition of atomic structure. There can be no unity-less, all-opposing forces, or completely opposition-less, all-unity of atomic structure. Complete unity im plies that the atomic structure is made of completely homogeneous stuff, with no opposing tendencies, while complete opposition implies making absolute the opposing tendencies at the expense of denying the existence of any unity whatsoever. For example, an atomic structure made completely of either positive charges or negative charges at the exclusion of the other, would elucidate the above explanation. The existence of one, in atomic, mo lecular, and macro-body structures, without the presence of the other is inconceivable.

F. Engels vividly clarifies this point:

All motion consists in the interplay of attraction and repul sion. Motion, however, is only possible when each individ ual attraction is compensated by a corresponding repulsion somewhere.' ... Dialectics has proved from the results of our experiences of nature, so far, that all polar *opposites, in general,, are determined by the mutual action of the two opposite poles on one another,* that the separation and opposi tion of these poles *exists only within their unity and interaction,, and, conversely, that their inter-connection exists only in their opposition.* This once established. there can be no ques tion of a final cancelling out of repulsion and attraction or of a final partition between one form of motion in one half of matter (atoms, molecules and bigger bodies), and the other form in the other half, consequently, there can be *no question of mutual penetration or of absolute separation of the two poles?* ..·· All natural processes are two-sided, ·they rest on the relation of at least two effective parts; action **and reaction.**:i

Engels goes on to comment on the gravitation theory of his time in the following manner:

The whole gravitation theory rests on saying that attraction. is the essence of matter. This is necessarily false. Where there is attraction, it must be compensated by repulsion. Hence, Hegel is quite right in saying that the essence of matter is attraction and repulsion. •

2.-Universality and Relativity of Unity and Opposing Tendencies of Self-Motioned Atoms

If we take the structure of self-motioned atom as the basis of our argument, which is in line with quantum mechanics, that is, investigating the material universe on an atomic and molecular level, or what Engels calls the theory of matter, we notice that, as previously explained, there is a duality of character associated with every self-motioned atomic structure. On the one hand, there are different atomic· and even· sub-atomic particles of opposing tendencies, constituting the atomic structure, and yet, these op posing particles can, as Engels confirms, only exist and live through an infinite number of "unified" processes. There is never a time in which these self-motioned atomic and sub-atomic particles live separately, on a non-unified basis. Therefore, unity and opposing tendencies are both simultaneously of universal· and relative character of every self-motioned atom, molecule, and macro-body. The unity and opposing tendencies and the degree they manifest themselves in each moment of existence depends, among other things, upon the level of evolution and development of the object we are talking about. .

In what sense are unity and opposing tendencies of self motioned atoms universal, and in what sense relative characters?

It was stated before that self-motioned atomic structure is made up of particles of opposing tendencies, and that opposite particles can only exist in unified processes, and that there is never a time in which a self-motioned atomic structure, regardless of its structural complexity and level of development, is either completely reduced to all-opposing tendencies or completely all unity posture. There is neither a completely unity-less, nor a completely opposing-tendencies-Jess self-motioned atomic struc ture. So in the sense that no self-motioned atomic structure can live while containing these dual characteristics, to that extent, unity and opposing tendencies are of universal, and as some author prefers to call them, of "absolute" character. By "absolute," I mean nothing else than the fact that their presence is necessary in order to make up the atomic structure. Or to put it more simply, they are universally required. But then individual self-motioned atoms do not all have the same degree of interactions between their atomic and sub-atomic particles; nor do they have com pletely the same unified processes; nor do they have a fixed, im penetrable relationship with other self-motioned atoms, especially, those adjacent to them.

In every stage of universal evolution, the self-motioned atomic universe *in parts, as well as "the whole," is undergoing, through a given, yet moving and constantly changing, unity-op posing-tendencies relationship.* The unified-opposing-tendencies of self-motioned atoms would be actively interacting, interpenetrating, and inter-conditioning out of which a different unity-opposing-ten dencies relationship emerges and develops. So you see, the only thing real in real life is change. Illsions arise when we take aspects of life as being fixed, eternally the same. Life rejects this entirely; bringing you a whole lot of disappointments. Therefore, seen from this point of view, both unity and opposing tendencies of self-mo-. tioned atomic structure are transitory, relative, and conditional, *because the processes of evolution would force them to change as opposed to their previous level of existence,* on a physical, chemi cal, and biological basis.

They are transitory and relative because they are going through a qualitative transition to either a higher level or decom pose to a different self-motioned atomic structure. But in the process, both the unity (what keeps them together, or attraction of the component parts), on the one hand, and opposing-tendencies (what keeps them apart, repulsion, "struggle") on the other hand, yes, both, *are changing relative to what they had been before.* So, the fact that both are changing as opposed to what they have been, *makes them relative, mutually conditional, and transitory. This discloses the relativity of their existence;* whereas, the fact that their presence is *permanently necessary* in any atomic, molecular and macro-bodily compositions or relationships makes them uni versal or of "absolute" character.

Here relativity of existence means relativity of matter and mo tion, or relativity of primarily self-moved entities and their relation ship to the general environment. Once again we find it necessary to. repeat a statement from Einstein's Theory of Relativity, for it ex presses this point very well.'

A. Einstein holds that:

The name "Theory of Relativity" is connected with the fact that motion from the point of view of possible experience, always appears as relative motion of one object with respect to another (e.g., of a car. with. respect to the ground, or the earth with respect to the sun and the fixed stars). Motion is never observable "as motion with respect to space" or, as it has been expressed, as "absolute motion." The principle of "relativity," in its widest sense, is contained in the state ment: The totality of physical phenomena is of such character that it gives no basis for the introduction of the concept of "absolute motion"; or shorter but less precise: There is no **"absolute motion."5**

Here Einstein refers to the relativity of "mechanical" motion of one given object with respect to other objects, and does not deal with motion on an atomic and molecular level and their mutual dependency; but, nevertheless, he expresses the relativity of motion on any level period.

One should not. politicize this statement and, therefore, based upon the above reasoning, conclude that the presence of both work ing-class and capitalist is permanently necessary for social progress.

The concept, as put forth by Lenin, describing the relation ship between unity and opposing-tendencies is not, in my opinion, in. the best tradition of Engels' teachings.

Lenin states that:

The unity (coincidence, identity, equal action) of opposites is conditional, temporary, transitory, and relative. The struggle of mutually exclusive opposites is absolute.[0]

Engels takes a more balanced approach. In a letter to P. L. Lavro, dated 12-17 November, 1875, Engels wrote:

... Of the Darwinian doctrine I accept the theory of evolu tion but Darwin's method of proof (struggle for life, natural selection) I consider only a first, provisional, imperfect ex pression of a newly discovered fact. Until Darwin's time, the very people who now see everywhere only *struggle* for ex istence (Vogt, Buchner, Moleschott, etc.) emphasized precisely *co-operation* in organic nature, the fact that the vegetable kingdom supplies oxygen an nutriment to the animal kingdom and, conversely, the animal kingdom supplies plants with carbonic acid and manure, which was particularly stressed by Liebig. *Both conceptions are justified, within cer tain limits, the one is as one-sided and narrow-minded as the other.* The interaction of *bodies in nature, animate as well as inanimate, includes both harmony and collision, struggle and cooperation.*"[7]

On this, Lenin underplays the significance of unity while overemphasizing the struggle. Engels sees the two as the necessary component parts of the same process, both being simultaneously relative, conditional, temporary, transitional, and of universal or, as Lenin calls it, "absolute" nature.

3- THE LAW OF QUALITATIVE TRANSFORMATION OF SELF-MOTIONED ATOMS

In the first law of evolution of the self-motioned atomic nature, the law of unified and opposing interpenetration of self motioned atoms, it was demonstrated that evolutionary processes require, as Engels observes, at least the active interaction of two entities or, as I like to put it, the dynamic interaction of the com ponent parts of self-motioned atomic structure, and that every self-motioned atomic composition is accompanied by a certain amount of matter (atomic and sub-atomic particles) and motion (self-motioned interactions) between them.

It is interesting to know that the number of atomic and sub atomic particles and their motions are not fixed and that, indeed, these particles have the ability, not necessarily of the same pro pensity, of transforming into one another, combining, and re combining with adjacent atoms, and therefore, causing a change in the quality of other self-motioned atomic structures involved. Soviet scientist, V. Rydnik maintains that:

... Up to now, particle ·transformations have dealt only with the electron (and, of ·course, the positron). After the discovery of the neutron, it was found· that it, too, is capable of transformation, but unlike the electron, not into field quanta but into other particles.•

If we start our argument by paying attention to atomic struc tures, building up our discussion all the way to molecular and macro-body level, then we see that, for example, the atomic struc tures of well over 130. known elements, such as, oxygen, hydrogen, and ozone, just to mention ·a few, are qualitatively different. For example, it is common knowledge that water is made up of two qualitatively different self-motioned atomic structures, or elements as they are called: oxygen and hydrogen. Both, being qualitatively different from each other, combine to form another still qualita tively different stuff, namely, water, which is different from its in dividual· parts from which it was made. Water can be decomposed into oxygen and hydrogen, self-motioned atoms.

F. Engels illuminates on the subject:

The *molecule is decomposed into its separate atoms which have quite different properties* from those of the molecule. In the case of molecules composed of various chemical elements, atoms· or molecules of these elements themselves make their appearance in the place of the compound molecule; in the case of molecules of elements, the free atoms appear, which exert quite distinct qualitative effects. The· free atoms of nascent oxygen are easily able to effect what the atoms of atmospheric oxygen,. bound together in the molecule, can never achieve." ... But the *molecule is also qualitatively dif ferent from the mass of the body* to which it belongs. It can carry out movements independently of this mass and while the latter remain, apparently,· at rest.[10] ·••• *The mass consists solely of molecules, but it is something essentially different from the molecules, just as the latter is different from the atom.11*

So we see that for something to change and become some thing else, which is to say, something qualitatively different, it needs to either give out some of its self-motioned atoms, its com ponent parts, or receive some. The process of "give and take" is very important in the working of nature. In the case of water, two different atomic compositions have to combine in order to produce something else, water. Therefore, any change of qualitative nature. must always be accompanied by one, or a series of, other self-mo tioned quantities.

F. Engels goes on to add:

All qualitative differences in nature rest on differences of chemical composition of or different quantities or forms of motion (energy) or, as is almost always the case, both. [Here, Engels observes the integrity and inseparability of matter and motion in the process of qualitative change.] ... Hence, it is impossible, [he continues,] to alter the quality of a body without addition or subtraction of matter *(opposing self-mo tioned sub-atomic, atomic or molecular particles)* and motion.[12] *(Various degree of complexity of interaction be tween them.)*

The underlined, clarifying statements in parenthesis are my own. Still continuing, Engels adds:

. . . Again, one can take the various proportions in. which oxygen combines with nitrogen, and sulphur, each of which produces a substance qualitatively different from the others! How laughing gas (nitrogen monoxide N 20) is from nitric anhydride (nitrogen pentoxide, N2 05) ! The first is a gas, the second at ordinary temperatures, a solid crystalline sub stance. And yet the whole difference in composition is that the second contains five times as much oxygen as the first, and between the two of them are three more oxides of nitrogen (No, N2, 03, N02), each of which is qualitatively different from the first two and each other.'"

A chemist can at will combine desired proportions of different elements and produce a new phenomenon. The existence of all available medicines testify to the validity of this law. But, in nature, self-motioned atoms bring about different things through the pro cesses of natural evolution.

Since the self-motioned atomic universe is constantly chang ing, evolving, and transforming, then everything, through inter action with the rest of the self-motioned material universe, undergoes a new quality. A change can only be born through involved interactions. "Anybody can," says Engels, "be virtuous (remaining pure) by himself, for vices (to produce something else, even though it may be corrupt) two are always necessary. Change of motion is always a process that takes place between at least two bodies."[14] Clarifying statements in the parenthesis are my own.

4.-PARTICULARITY OF QUALITATIVE TRANSFORMATION OF VARIOUS PARTS OF SELF-MOTIONED ATOMIC UNIVERSE

Not all parts of self-motioned atomic universe are transform ing, and evolving at the same pace, with the same rapidity, and exactly the same forms. Nature, in that, is very creative in develop ing infinite forms, some already existing, others having become extinct, while many will be emerging. This is despite the fact that evolution is a universal phenomenon. But on what are different forms of qualitative transition based? On what is each specific form· of transformation based?

The evolutionary qualitative transformations depend upon the complexity of self-motioned phenomena, what they have gone through in terms of evolutionary development, and how they in teract with the rest <>f self-motioned atomic nature, and its corre sponding level of development. The more complex a given aspect of nature, the slower and more complicated the process of qualita tive transformation.

For example, the manner in which biological organisms, hu man beings, are qualitatively transforming, is quite different from other self-motioned atoms with lower levels of development, i.e., vegetation. And the differences lie upoa the molecular, geneti_c, and cellular complexities of the entities involved. The particular ities and specificities of qualitative transformations should not be absolutified for indeed, it would . be at the expense of the commonality of all· self-motioned atoms. Just as ignoring the par ticularities of qualitative transformations of various parts of self-motioned atomic nature would result in seeing nature. as completely uniform, *something which it is not.* A complex process evolves not in· one stroke, but in a very complicated gradual manner. The process acquires the characteristics of what the entity will become, in a piecemeal form, where the more basic foundation of evolution would be gradually realized.

The following is a specific application of the law of qualitative transition in a human love affair. Suppose you love a person of the opposite sex very deeply. Suppose that your love is charac terized by your heart being completely, symbolically, that is, spot less, meaning your entire heart belongs to that person. Suppose further that your heart, expressing the gravity of your love, is equivalent to a given quality. You continue interacting with that person. You do your darn best to keep that individual happy. But, the more you··try to please her/him, in return, the more vicious, rebellious and diabolically disrespectful toward you she/he be- comes. Every time you go through a very unpleasant experience I with her/him, a bruised spot is recorded in your heart. To convert :1 your love, a given quality, to hatred of that person, another quality, your heart needs to be entirely bruised. But bruising the entire heart[1]

is not done at one stroke. It is gradual and piecemeal. It keeps I receiving bruises until such a time where the entire heart is bruised. When this happens, you hate that person instead of loving her/him. But this new quality, the hatred, was gradually realized until it be- came a· full-fledged monster. This shows the thin and overlapping wall between love and hatred, and how one becomes the other.

· Take another example, showing a specific application of the law of qualitative transformation, as related to another aspect of human life. Suppose you plan to go to college to specialize in a given discipline, say, becoming a physician. Obviously, after having been investigating the· subject matter for a number of years, and having received· a degree indicating that you learned the discipline reasonably well, you are supposed to have sufficient knowledge. of your profession. Suppose that the degree of your medical informa tion, in its entirely represents a given quality. But, remember all the hours, days, months and years you sat through classes, listening to your professors. Every hour of your investigation, if we could consider an hour as a unit, represents a part or portion of the entire medical knowledge you possess. Certainly, in every moment of your sitting through discussions, a degree of your knowledge was acquired. These moments of knowledge kept accumulating. Every moment of the accumulation of this knowledge, this quality, stands as an active partial realization of your total medical knowledge, which is the quality we have talked about. The accumulation of partial bits of knowledge eventually resulted in an organically connected body of knowledge, a new quality.

There is a similarity between the first and the second ex amples. In the first example, the new qualitative transformation to materialize was to get the whole heart being totally bruised. Whereas, in the second example, possession of reasonable degree of medical knowledge was a qualitative transition. In both cases, the qualitative transformations had to be accompanied with bits of quantitative additions, the accumulating and interaction of which, gave birth to emerging qualities.

The third law of universal development, as formulated by Engels, is the law of negation of negation, or what I call: The law of negation of negation of self-motioned atoms.

5.-THE LAW OF NEGATION OF NEGATION OF SELF-MOTIONED ATOMS

Universal change and evolution of self-motioned atomic existence presupposes that every individual, self-motioned atom "is" and "is not" at the same time. It "is" because it represents certain relative patterns of stability in terms of material organization at one given moment. It "is not" because its future moments of exis tence are not exactly a mere duplication or repitition of what a self motioned atom, self-motioned molecule, or self-motioned macro body are, say, at this moment.

All these complicated moments of existence, when combined, represent a change, a "new quality." That means what a self-mo tioned atom, molecule, or macro-body "will be" in the future, would not be the same as what they "are" today. For instance, let us take the example of moments of existence as related to the life of an individual from the moment he is conceived up to the moment he dies. Say that moments of his entire life can be represented by seconds, something that we could measure in terms of time. We could approximately figure out how many seconds there are in his life. Notice the continuation of his life from the time where he is nothing other than a combination of his parents' sperm and egg interacting, preparing the conditions to form an embryo, up to be coming a fetus in a few months, at which time it would begin' to develop relatively distinct features, up to the time when he is born, going through childhood, ado-lescence, adulthood, old age, and, eventually, death. We could divide his ·entire life into moments, "seconds," of existence. We could then see that every moment or "second" of his life is relatively different from previous mo ments. I said "relatively different" because of the relative con tinuity of the interrelated moments of his life which, combined, represent a *creative continuity.*

Had every moment of his life been exactly the same as the previous one it would have always remained what it originally had been; a combina-tion of his parents' sperm and egg never going through the evolving stages it did. We see the continuity and dis continuity constitute a very creative process. All the succeeding moments of his life are creatively continuous and discontinuous at the same time.

Let us now go back to our main line of argument; evolution of self-motioned atoms. We said that moments of existence of self-motioned at-oms would result in a change, a "new quality." We argued that even though there is continuity among all the moments, nevertheless, the succeeding moments are relatively different.

But what is important is the nature of this difference and change exhibit-ed by moments of existence. What is the nature of this change? Is the change of "complete nature?" In other words, what is the relationship between what a self-motioned atom and, **for that matter, anything in the universe,** "was," **"is," and "will** be," the relationship between the past, the present, and the future? In what way are these stages, through which all different self-motioned atoms necessarily go, related? Can a self-motioned atom be "totally different" than what it had been in the past? Or will it ever become "totally different" from what it "is" at this very moment? The concept of "total replacement," or "total change" is false and implies that different stages of an object's development are unrelated and that something can become "totally" something else. The concepts of "totally," "absolutely," "entirely," and "ultimately" are nothing other than cancerous diseases, eating off of human brains, causing confusion and a state of mind that is self-destructive as well as being dangerous to others. The mentality that insists on "totally," "absolutely," "entirely," and "ultimately" is the one that does not understand interrelationships but, instead, freezes life into unrelated individual parts, elevating them to holy, absolute distinctions.

F. Engels, talking about the evolution of human societies, from a feudal to a capitalist and transition to a socialist one, and the corresponding moralities that the social classes, such as the feudal aristocracy, the bourgeoisie, and the working class, upon their emergence, develop, and the relationship of these moralities, had the allowing to say:

But when we see that the ·three classes of modern society; the feudal aristocracy, the bourgeoisie and the proletariat each have a morality of their own we can only draw the one conclusion: That men, consciously or unconsciously, derive their ethical ideas in the last resort from the practical rela tions on which their class position is based-from the economic relations in which they carry on production and exchange. But, nevertheless, there is quite a lot of which the three moral theories *have in common-is* this not at least a portion of a morality which is fixed, once and for all? *These moral theories represent three different stages of the same historical develop ment, having, therefore, a common historical background, and for that reason alone they have much in common ...*[10]

This is a beautiful example of morality seen in the process of evolutionary transition, in which morality of a specific kind comes into being, not in an abstract form, but rather, by requirements of real life as practiced by the social classes, or to put it more ac curately, as the way they relate to relations of production and exchange, and not based upon the sermons the ministers deliver on Sundays, all day. And what is more important is that when you see things in an evolutionary way, you are capable of observing the similarities and distinctions of different stages of. development. And that is exactly what Engels has done.

That is why we must see the socialist experiences as practiced in the Soviet Union, Eastern Europe, and in China not as some thing being "totally" different, but rather as the further evolution of the same historical background, all of them having been born out of the capitalist womb, with all its imperfections.

Self-motioned atoms create the conditions for becoming changed, evolved, and further developed. When this process takes place, we say that one stage of development of self-motioned atoms is being replaced (negated) by another stage. Here nega tion does not mean anything more than "one stage has changed, evolved, and developed." Since what "is" can not always remain **"is," and that it has got to become "will be," then what "is" will** necessarily become "will be." The past of the self-motioned ob jects was evolved (negated) by the present; and sure enough, the present will be, by necessity, changed, evolved (negated), and developed. What was negated is further negated, and what is negated will be further negated. Any development is further de veloped. This is the reasoning behind the use of the term "nega tion of negation." One could creatively apply this universally true law to one's own experiences of any kind.

1. F. Engels, *Dialectics of Nature,* p. 38.
2. F. Engels. *Dialectics of Natlure,* pp. 38-39.
3. Ibid., p. 51.
4. Ibid., p. 259.
5. Albert Einstein, *Out of My Later Years,* p. 41.
6. V. Lenin, *Philosophical Note Book,* on the question of dialectics, pp. 359-369.
7. Dilip Bose, *Society. And Revolution, Essays in Honour of Engels Dialectical Materialism and Science* (New Delhi, Ahmedabad, Bombay: People's Publishing House), p.l86.
8. V. Rydnik, *ABC's of Quantum Mechanics,* pp. 277-278.
9. Engels1 *Dialectics of Nature,* p. 28.
10. Ibid.

11. Ibid., p. 29.
12. Ibid., p. 27.
13. Ibid., p. 31.
14. Ibid., p. 28.
!5. F. Engels, *Anti-Duhring*, p. 114.

CHAPTER 8

Causality-Cause and Effect

THE CONCEPT OF CAUSALITY (THE relationship between cause and effect) must also be considered from the position of self-motioned atom, complex self-motioned atoms (molecular level), complex self-motioned molecules (more evolved bodies).

I.-SELF-MOTIONED ATOM, BEING *PRIMARILY* SELF-CAUSED, SECONDARILY ENGAGES IN MUTUALLY DETERMINED CAUSATIONS.

Engels's views on inter-self-motioned atomic, inter-molecular, inter-macro-body causality, and their mutual determinism are of great significance: "The first thing that strikes us in considering matter in motion *is the interconnection of the individual motions of separate bodies, their being determined by one another.'*

2.-THE RELATIONSHIP BETWEEN SELF-MOTIONED ATOMS AND THE REST OF THE MATERIAL UNIVERSE.

Every individual self-motioned atom is primarily self-moved and self-caused. That simply *means that movements and motions associated with individual self-motioned atoms are internally, creatively-law-governed.* However, every individual self-motioned atom, while having its own primary motive forces and its own causal relations, is a condition for the development and further evolution of other self-motioned atoms, molecules and macro bodies. So every self-motioned atom is *primarily* self-propelled, self-moved, and self-caused, being simultaneously a causally con tributing factor, joining with other self-motioned atoms, molecules and macro-bodies to bring about (cause), give birth to other self motioned entities and eventually causing the "whole."

That simply means that there is no fixed and once-and-for-all ***whole"*** ***or universal causal relation, and that the whole1 is being*** creatively born *and reborn out of individual self-motioned causa tions and their ever-evolving interactions.* More simply put, the individual self-motioned atoms and mol-ecules are *primarily* self moved and self-caused, while they, *secondarily,* join to cause, bring about, give birth to, a different complex phenomenon.

A cause can be jointly produced; therefore, we have to talk about self-motioned, inter-atomic, inter-molecular, and inter macro-body causality, where every self-motioned entity is contrib uting to a given material situa-tion, which results in the birth of a new self-motioned body, where indi-vidual causes and effects are interacting, jointly causing and generating a universal causality or causality of the "whole." In return, once the "whole" is in active interrelated motions, it influences, conditions the direction of evolutionary processes of individual self-motioned atomic, molec ular, and macro-body causations. There develops an active and mutually de-termining highly complicated causal interrelationship between individual self-motioned atoms and the general environ ment.

F. Engels clarifies this causal interrelationship:

Further, we find upon closer investigation that the two poles of an antithesis, e.g., positive and negative, are as inseparable as they are opposed and that *despite their opposition they mutually interpenetrate.* And we find, in like manner, that cause and effect are conceptions which only hold good in their application to individual cases, *but as soon as we con sider the particular cases in their general connection with the universe as a whole, they run into each other and they be come confounded* [mingled so that individual, self-motioned entities cannot be individually distinguished], when we con template that universal action and reaction, in which causes and effects are eternally changing places, so that what is in effect here and now, will be a cause there and then?

Not only does our group of planets move about the sun and our sun within our island universe, but our whole island universe also moves in space in temporary, relative equilib rium with other island universes, for *even the relative equilibrium of freely moving bodies can only exist where the motion is reciprocally determined."*

The clarifying statement in parenthesis is mine.

Take the example of modern mass production. Every individ ual worker, whether physical or mental, enters a given production facility, say, General Motors, with certain skills, know-how, educa tion, and training which are needed for one or several aspects of production processes. To that extent, all of these productive quali ties together, potentially form a degree of individual self-motion, self-movement, and self-initiative which *cannot become assertive in terms* of producing material goods and services until the entire plant is organized according to a productive and efficient plan, in which case the creative interaction and interplay of all individual, *primarily self-motioned workers* results, causes, or brings about a given outcome; the cars we drive. This is a clear example of in dividual self-motioned causes (the workers) engaging in collective activities, out of which a certain thing is produced. *Primarily self moved workers* combine their force creatively to *secondarily, col lectively,.* cause (produce) the cars.

Take a soccer game. It is now becoming popular in the United States. Every player is self-moved, that is, his movements are primarily self-caused. He can do certain things, take certain self-initiated activities, such as, taking, controlling, and maneuver ing the ball through the "authorized" field. For this he does not have to have the cooperation of the other players. But then, this . is not a game as of yet. The game begins when all the individual, self-motioned, self-caused soccer players are set into a very dynamic and complicated motion, where the individual move movement of the *players are actively modified by the dynamic interaction of the players.* The aim of all soccer players *is* to make as many goals as they can. This requires the conscious dynamically coordinated activities of self-motioned, self-caused players of one team interacting with those of the other team. It is out of this *primarily individually-self-moved and secondary, and collectively interacting that the entire game is born, in which certain effect, such as making goals is dynamically generated.* The effect, making goals, is an example in which it was born, caused by collective contribution of *relatively independent and primarily self-moved and self-caused soccer players.*

Here the entire process and, as Engels calls *"the motion, is reciprocally determined."* As you see, the intended aims of all players can only find embodiment in *real interaction* in which the results may be quite different from those intended. The result, the game-in "totality," the universal causality-the game, was dynam ically born out of interaction of self-motioned individual acts.

Take another example: A capitalist economy is based upon private ownership of business enterprises. Every enterprise is run by the capitalists themselves or their hired managements. Every enterprise is primarily self-moved and self-caused, That means they initiate their own plan as to how much capital to employ, and where, to be combined with what number of workers, and, eventually, what to produce. All of these decisions are self-made, some even without consideration of otherwise better-coordinated needs and requirements of the society as a whole. Once all of these individual business enterprises are set into motion, interact ing, there emerges something basically unintended by the individual enterprises; unemployment, overproduction of certain com modities and underproduction of others, over-expansion of massage parlors and negligent under-production of health care facilities for the entire population, over-pollution of the environment, oceans, and rivers, conducting wars of aggression, a generation of econom ically-determined corruption, such as, the multinational corpora tions' bribery system to generate business and to control the economic life of people on the international basis.

All of these universal outcomes and causes were produced out of the interaction of the self-moved enterprises and their self oriented economic interests.

Bnt once you have a universal posture of the economy as explained, it is not a static situation. The economy as a whole actively leaves an impact upon the individual enterprises. The overall overproduction and a corresponding lack of market to sell the overproduced commodities would compel the individual en

terprises to start laying off workers if they do not succeed in dumping their overproduced goods in foreign market The fact that at this writing 10 percent of the American labor force is out of jobs, testifies to the validity of this statement. And if the over produced commodities cannot be dumped abroad, then that calls for laying more people off, resulting in increased unemployment lines. Many small businesses which do not have very much capital, and cannot afford to incur losses until such time as things would get better, would go bankrupt. The universal over-pollution would start damaging the health of the people of society, with the possibility of causing undesirable mutations. The unemployment among women would increase, the mushrooming of massage parlors, which are, in reality, a form of prostitution-American style-drawing more of our sisters, wives, and mothers into the prostitution business. Accompanied with this, is the emergence of layers of pimps, professional thieves, and a well-organized drug traffic to keep these people "entertained." All of these combined would increase corruption, debase our morals, develop calluses in our hearts and minds.

Once these become a way of life, a good portion of the society becomes criminally oriented on a social basis; biologically, they become carriers of all kinds of diseases, passed to the next genera tion. The argument is that they should have the right to destroy themselves. But the story does not end here because the diseased and destroyed bodies *can only give birth to offspring who are genetically diseased and physically deformed.* These would be their gifts to succeeding generations.

Take another example of modem molecular biology. Every trait of a biological organism is said to be attributable to one or a group of genes. Every gene has its own *primarily* self-moved, self caused characteristics and activities with a given evolutionary posture. Yet it would be absurd to absolutify the biological im portance of an individual gene at the expense of the collective behavior of all the genes in maintenance and the evolution of organisms. Taking another *extreme point of* view *would be to ignore the contribution of individual genes,* their relative existence, movement, functions, and relatively unique participation in evo lutionary processes; while emphasizing the "whole" in an abstract manner. The argument between molecular biologists and organis mal biologists explains the two current tendencies in biology. The molecular biologists maintain that a biological organism, say, a human being, is, literally, made up of millions of individual genes out of which the entire body is put together. The understanding of the makeup, behavior, and functions of these genes would "reveal the secrets of life." The organismal biologists hold that biological organisms cannot be understood by dividing the organism into individual pieces and their corresponding study. The individual genes can only make sense while interacting as an "unit," the organism as a whole.

It seems to me that without understanding life on a molecular and genetic level, it would be rather impossible to understand the nature of life as a "whole" unit.

But all of these wonderful molecules and genes can only assert themselves in a collective and biologically organized man ner. To absolutify either of these two tendencies at the expense of the other would be very damaging to the development of biology as a science. There is so much that these two tendencies can learn from one another.

George Gaylord Simpson's statements on the issue are very timely:

Wald has said that, "Living organisms are the greatly magnified expression of the molecules that compose them." Anfinson believes that "we may almost define the life sciences as those concerned with elucidation of the mecha nisms by which *molecules exert their specific actions in living cells* ..." Still continuing, he goes on to say that: "Everyone agrees that eventual understanding of *relationship between levels is necessary* ..."Weiss has pointed out that there is a cellular control of molecular activities, as well as molecular control of cellular activities. There is also an organismal control of cellular activities and, for that matter, populational control of organismal activities.• Finally, he concludes by saying that: "Living systems, that is, organisms, are incom parably more complex. To study these systems it is possible and necessary to analyze them into simpler components, for example, to study intercellular chemistry one reaction at a time. However, the system, as such, that is, that the life of the organism, cannot be understood in terms of any or all such relations taken one at a time but only considering all at once within the structural and chemical system as a whole." •

George G. Simpson talks about the significance of relative existence, operation, and functions of biological organisms on molecular, cellular, organismal, and populational levels and their interrelationship in terms of *mutually determining* (controlling, influencing, and conditioning) one another in the process of evolution.

Engels's remarks, pointed out a hundred years earlier, pro foundly testifies to this very important fact on a much broader basis:

Reciprocal action is the first thing that we encounter when we consider matter in motion as a whole from the standpoint of modern natural science. We see a series of forms of motion, mechanical motion, heat, light, electricity, magnetism, chemical union and decomposition, transitions of states of aggregation, organic life, all of which, if at present we still make an exception of organic life, *pass into one another, mutually determine one another, are in one place, cause, and, in another, effect* ...0

1. **Engels,** ***Dialectics of Nature,*** **pp. 170-171.**
2. **F. Engels,** ***Anti-Duhring,*** **p. 32.**
3. **F. Engels.** ***D***
4. George Gaylord Simpson, *Biology and Man* (New York: Harcourt, Brace, Jovanovich,1nc., 1969),·pp. 33 34.
5. Ibid., pp. 33-34.
6. Engels, *Dialectics of Nature* p. 175.

CHAPTER 9

Truth

———

FORCED TO COME UP WITH certain answers relating to our daily problems, we spend so much of our time every day arguing whether something is, or is not, true. For example, what is the truth of how life started? Is it true that 7,000 :years ago human beings were created? Or is it true that human beings evolved out of lower forms of life, namely, the ape-and that both of these forms of life, as well as other biological organisms, share com mon ancestors? Is it true that in a capitalist society workers are exploited by the capitalists? Is it true that one given human race or even one type of animal may be more intelligent than other races or animals, respectively? We could, literally, formulate mil lions of questions relating to any particular part of life and pose the question of "is it or is it not true," that certain things are in certain ways. What do all of these questions have in common? Is there any way of establishing their truths? Or is it a hopeless case that we should give up as soon as we possibly can and forget about it?

So far we have become acquainted with an evolutionary logic, namely, self-motioned atomic dialectical materialist logic. Using the principles of this logic, we can arrive at understanding the truths of our material universe and whatever it contains.

Going back to our line of questioning about the truths of everything, what they have in common is that. we have infinite number of material things in life, i.e., .vegetables you may grow in your backyard, the flow-ers· you may receive for Valentine's Day (regardless of whether or not you deserve it), your pet at your home who keeps you company, just to men-tion a few. At the same time, we have ideas, explanations, theories, and a general understanding as to what these things are, how they behave, how they function, what they are made of, what makes them tick, what they are being used for, and so forth.

When we have a medical problem, we do not go to the medicine cabinet and just randomly pick up any one bottle or jar that we can reach first. We would carefully read each label and find out exactly which par-ticular medicine is good for our particu lar malady. The label on the· medicine jar must correspond to what the medicine is chemically com-posed of, what it is good for, and a lot of other related questions. So, what we are trying to do when we ask "is it or is it not true," is to establish whether. or not what we have in mind about these material things, i.e., ideas, impressions, understandings, thoughts, theories, and conceptions, *really correspond to the actual state of beings of these things.* In short, we are trying to establish the truths about them, just as the description on the medicine bottle, the chemist's idea, cor responds to what the nature of that medicine is. All ideas must then correspond to, and reflect, the material life as closely as pos sible. Therefore, in order to establish the truths of some thing, the problem of correspondence and reflection be-come very important.

V. Afanasyev, in his book *Marxist Philosophy,* states the Marxist doc-trine of truth, as follows:

Dialectical materialism understands of truths as that knowledge of an object, *which correctly reflects this object,* i.e., *corresponds to it.* For ex-ample, the scientific proposi tions that "bodies consist of atoms," that "the Earth existed prior to man," that "people are makers of history," etc., are true.[1]

1.-Simple Reflection and Correspondence of Material Universe is Not Enough to Establish Truth: An Evolutionary Understanding Is Needed.

Here, Afanasyev's understanding of truth is based upon whether or not our knowledge of something does or does not, partially or completely, reflect that object. And if there is com plete reflection and correspondence between our knowledge, on the one hand, and the object, on the other hand, then we are, supposed to have establish truths about that object. However, if there is no complete correspondence and reflection, then we may have falsehood, half-truth, or even, complete untruth.

The problem of establishing truths as related to a material universe, in my opinion, is twofold: First, we have to understand an object, its internal structure, its movements, its connections, its relationship with the rest of self-motioned material entities, its role in the self-evolutionary, as well as universal evolutionary processes. In short, the material universe must be seen in move ment. Truth, reflecting this evolving universe, must be dynamic. After we have established an evolutionary understanding of a self-motioned atomic universe, then we may ask whether or not our knowledge reflects, and corresponds to an object and is, **therefore, true."**

To reduce the problem of truth to simple mechanical cor respondence at tbe expense of failing to understand the nature of an object in movement, is inadequate in establishing truths from the standpoint of self-motioned atomic dialectical materialist logic.

Now, let us attempt to analyze some of the problems as sociated with establishing truths of the material universe.

2 -EVOLUTIONARY SELF-MOTIONED ATOMIC UNIVERSE AS THE BASIS OF OBJECTIVE TRUTH

First, let us show a degree of consistency in considering the self-motioned atomic structures as the point of departure, in establishing truths, even at the expense of some redundancy.

In our previous discussion, we arrived at the understanding that even though an atom is primarily self-moved, it does not have an absolutely isolated existence from the rest of the self motioned atomic universe; and that it actively engages in the pro cesses of self, as well as, universal development. This, in my opinion, invalidates the argument of "absolute," or total autonomy of individual self-motioned atomic existence and, for that matter, isolated, absolute isolation of anything.

Truths *about the self-motioned atomic universe must be ex tracted from material life and not fabricated in our mind and imposed upon it.* Therefore, truth is objective. It is derived from the material objective life.

Let us, for the purpose of clarity, attempt to see the material universe in two different forms: (1) seeing and investigating it as fixed, static, stationary, eternally the same, non-evolving fashion, and (2) analyzing it in movement, change, and interrelated evolutionary manner. Having done so, let us attempt to resolve the problem of reflection and correspondence of our ideas to material life, which are supposed to establish truths within the context of these two contrasting ways of seeing things.

3.-STATIONARY, NON-CHANGING, NON-EVOLVING MATERIAL UNIVERSE AND THE PROBLEMS OF ESTABLISHING TRUTHS

Let us assume that all individual atoms, molecules, and big ger complex material objects are totally isolated from one another, and that they are of non-changing character. Let us further as sume that bodies move, but their movement does not in any way change the nature of individual material things. This would be the kind ·Of material world we have. Our problem is to get as much knowledge (truths) about it as we can. The process of acquiring knowledge of this stationary, non-changing universe could be very easy. We would start accumulating knowledge (truth) until we would simply exhaust the process, at which time there would be nothing else left to know about everything. At that time whatever we know about the material universe would al legedly completely reflect and correspond to the material universe, and would, therefore, be the "absolute truth." For example, sup pose you have to travel from Los Angeles to San Francisco. Suppose further, that you start from one designated point in L.A. to another designated one in San Francisco. There is a def inite amount of mileage to be covered. Every time you complete a mile, you are closer to your destination in San Francisco. If you continue driving, you would reach a point where there is no more distance left and that you have exhausted the entire mileage distance. You reached your destination point "completely and en tirely" because you had supposedly two fixed points, one to start from, and the other to go to.

This analogy should convey the message, because establishing truths about a fixed material universe would only involve accumulating knowledge until such a point where there would be no more to know, just as in case of driving from Los Angeles to San Francisco, if we continue covering mileage, we would reach a point where there would be no more distance to cover. In both cases, we have two fixed points to reach, And, in both cases, we reach exhaustion points.!

4.-Self-Motioned Atomic Universe Is Not Fixed, ' Stationary; It Is Constantly Changing and Evolving.

The self-motioned atomic universe is not stationary, non changing, and non-evolving. Every atom is not only evolving, but is also connected to the immediate as well as the general material environment. That is why

Engels thought that dialectical material ist logic or what I have alternatively called self-motioned atomic dialectical materialist logic is to be developed "as a science of interconnections, in contrast to metaphysics."2 Here by "metaphys ics," he means an outlook that sees everything as being fixed and eternally the same.

On the importance of motions in the process of evolution and determination of the nature of different things, Engels goes on to say that:

The different forms and varieties of matter are themselves to be recognized through motion,. only in this are the properties of bodies exhibited; *of a body that does not move there is nothing to be said. Hence, the constitution of moving bodies results from the forms of motion.'*

·To Engels, understanding of material connections and inter connections is very important. Connections influence and finally help change a material situation. Just as if you want to know a person, in addition to finding out how he/she is, find out where he-she is connected to, i.e., the kind of friends he/she hangs around with. A person's personality is not formed in a vacuum. It is born in interactions with others.

Finally, emphasizing the significance of connections in the process of universal evolution, Engels concludes: "In the present work dialectics is conceived as the science of most general laws of motion";[4] "Dialectics as a science of total connections."• One can attempt to establish connections on all levels in our dally life, ranging from sub-atomic, self-motioned atomic, molecular, genetic, cellular, organismal, populational, psychological,. emotional, eco nomic, and environmental, human love affairs, just to men tion a few. These connections and interconnections do also change and evolve and, therefore, influence the nature of a material entity.

Connections can be conditions in which a material entity develops. For example, between the ages of twelve and sixteen, I was in a set of conditions that were very conducive to my be coming and remaining a rowdy hood-lum. There I became one-and a good one, too! At a later time I became involved in educated circles that knew the importance of self-improvement for the purpose of self-seeking interests. I was influenced enough by this group that I went to England to 'become an engineer. Upon my coming to the United States, I was placed in another set of conditions that talked about noble humanitarian ideas, the hope of man to live in a secure society in which one works for the betterment of the entire society when the entire so-ciety commits itself to material and spiritual security and enrichment of the one. Every one of the preceding situations offered a set of condi tions; but none was as comprehensive and as creative as the very last one. I responded so organically, as a sunflower plant follows the movement of the sun's radia-tions. It is fair to say that my mentality has been shaped in the background of these seemingly contradictory sets of conditions. The important thing to remember is that conditions and connections changed, and when they did, they influenced and changed my life. Who knows what follows? The idea of every material entity needing the material associations of others in order to further evolve, require us to deal with this very important issue and, there-fore, introduce the concept of *relativity of existence.*

5.-RELATIVITY OF EXISTENCE OF SELF-MOTIONED ATOMIC UNIVERSE

Referring to Afanasyev, truth is a "total reflection" of material bodies. But, Engels says: "of the bodies that do not move, there is nothing to be said." Bodies do not move exactly in similar manner al the time. They move and evolve creatively. Now then, what have we so far proved? *We have shown that every self-motioned atom, molecule, and complex self-motioned macro-body is relatively dependent upon the rest of the self-motioned atomic universe for its continued existence and development.* We, therefore, have to introduce the concept of *relativity of existence* as opposed to totally autonomous and isolated atomic existence. Relativity of material existence means nothing more than considering self-motioned atomic entities as having *relatively individualized existence,* never in isolation, but always maintaining active inter relationships with the outside world; providing conditions for other self-motioned entities, while receiving necessary and chance conditions for their development. *That puts the evolution ary processes on creative mutual determinism of all things, with the primary emphasis on the evolutionary, self-movement of dif ferent parts of the material universe.*

On relativity and reciprocal influence of self-motioned atomic existence, Engels had the following to say:

So long as we consider things as static and lifeless, each one by itself, alongside of and after each other, it is true that we do not run up against any contradiction *[inter related self motioned processes].* Within the limits of this sphere of thought, we can get along on the basis of the usual *metaphysi cal mode of thought [seeing the universe made up of un related, un-evolving individual things. But the position is quite different as soon as we consider things in their motion, their change, their life, their reciprocal influence on one another. Then we immediately become involved in contraradic tions. Motion itself is contradiction,* even simple mechanical change of place can only come about through a body at one and the same moment of time, being both in one place and in another place, being in one and the same place, and also not **in it.** *And the continuous assertion and· simultaneous solution of this contradiction is precisely what motion is.*[6]

Here, Engels employs the term, "contradiction," to mean process of motion. The clarifying statements in capitalized letters and in parenthesis are my own.

The processes of self-motion found **in** nature and life are never-ending. Therefore, "contradiction," which really means the same as processes of self-motion will always be present in different *forms.* To attempt to remove "contradictions" from life "once and for all" would be equivalent to desiring to remove processes of motion from matter, or material life.

Here Engels mocks the attempt and desire to remove the processes of self-motion (contradiction) from life by stating:

With all philosophers **it** is precisely the "system" which is perishable; and for the simple reason that it springs from an imperishable desire of the human *mind-the desire to over come all contradictions. But if all contradictions are once and for all disposed of, we shall have arrived at so-called absolute truth:* world history will be at an end ...'

There is a danger in seeing things as having isolated exis tence. And if this type of thinking creeps into philosophy, it will create tremendous ham to the further development of mankind. A case in point is the false sense of individualism in America, where the individual and his interests are *not* seen within· the context of an integrated whole, the entire society, but the individual in isolation, who can do "as he desires." This mode of thought, unfortunately, which is prevalent in natural sciences, philosophy, and social sciences and what is worse in our personal attitude in America, is the characteristic of the capitalist era.

While capitalist nations formulate their foreign policy based upon this mentality, every one of them moving from the position of self-interest only, often at the expense of others, they are bound to run into major conflicts and mutual suspicion, sometimes to be resolved through military means. A superficial look on the inter- national scene very easily testifies to this. However, the process of seeing everything in an integrated way in natural sciences, phi- losophy, social sciences, as well as personal behavior, with the desire of reconciling the interests of individual nations in a collec- tively coordinated and a beneficial manner, has already started and developed, to some extent, among the "Socialist Community."

Engels's remarks are very expressive of the point of view involved:

The *analysis of nature into its individual parts, the grouping of the different natural processes and natural objects in def inite classes,* the study of the internal anatomy of organic bodies in their manifold forms-these were the fundamental conditions of the gigantic strides in our knowledge of nature which have been made during the last 400 years. But this method of investigation has also left us as a legacy, the habit of observing natural objects and natural processes in their isolation, *detached from the whole vast interconnection of things; not as essentially changing, but as fixed constants, not in their life, but in their death. And when, as was the case with Bacon and Locke, this way of looking at things was transferred from natural science to philosophy, it produced the specific narrow-mindedness of the last centuries, the metaphysical mode of thought.* •

John Locke, whose intellectual and philosophical ideas not only shaped the beliefs of the American revolutionary forefathers, but also as a philosopher whose thinking, to this very day, still remains the basis of the American educational system and, there fore, American mentality maintains that matter, being firmly at rest, is a dead lump, incapable of producing any motion by itself. When he is gracious enough to admit that here and there motion could be found, he holds that "motion is added to matter by some other being more powerful than matter." Here is what the man has to say:

Let us suppose any parcel of matter eternal, great or small, we shall find it in itself *able to produce nothing.* For example: Let us suppose the matter of the next pebble we meet with, eternal, closely united, *and parts firmly* at rest together; if there were no other being in the world, must it not eternally remain so, a dead, inactive lump? Is it possible to conceive it can add motion to itself, being purely matter, or produce anything? Matter, then, by its own strength, cannot produce it itself so much; the motion it has must also be from eternity, or else be produced or added to matter by some other being more powerful than matter: matter, as is evident, having no power to produce motion it itself."9

This indeed is the epitome of ignorance and narrow-mindedness of the century that Engels talked about.

Mankind, through experience, is becoming conscious of the shortcomings and inadequacy of considering things as unrelated, finished, dead entities and, therefore, incorporating into its men tality self-motioned atomic dialectical materialist mode of thinking that sees the universe as an ever-evolving interrelated whole.

Maurice Cornforth, a British Marxist philosopher, employing a quotation from Engels, maintains:

The old rigid antitheses, the sharp impassible dividing lines are more and more disappearing ... The recognition that these antitheses and distinctions are in fact to be found in nature, but only with relative validity and that, *on the other hand, their imagined rigidity, and absoluteness have been introduced into nature only by our minds-this recognition is the kernel of dialectical conception of nature.* [10]

Joseph Dietzgen, who, according to Engels, arrived at dialec tical materialist logic independently of Marx, Engels, and even Hegel, a philosopher whose work is least studied, and whose authority on the subject matter remains, unfortunately, unrecog nized, comments on the relativity of self-motioned material ex istence as follows:

Every existence is relative, relates itself to others, and enters into different relations of time and space with them ...[11]

But the actual world is absolutely relative, absolutely tran sient, an infinity of manifestations, an unlimited quality. All are simply parts of this world, partial truths. Semblance and truth flow dialectically into one another like hard and soft, good and bad, right and wrong, but at the same time they remain different.'"

Finally, Engels, still talking about evolutionary relative material existence, and therefore, evolutionary thinking concludes:

... The great basic thought that the world is not to be com prehended as a *complex of ready-made things,* but as a *complex of processes,* in which the *things* apparently stable, no less than their mind-images in our heads, the concepts *go through uninterrupted change of coming into being and passing away, in which,* in *spite of all seeming* accidents and of all temporary retrogression, a progressive development asserts itself in the end-this great fundamental thought has, especially since the time of Hegel, so thoroughly permeated ordinary consciousness that in this generality it is scarcely ever contradicted.'"

Now, let us show relativity of material existence as found in the human body, just one aspect of material existence. My father had diabetes. His diabetic conditions began affecting his eyes, causing spillage of sugar in his blood, worsening his general condi tion. This, plus obesity and his high blood pressure caused the breakage of some of his eye blood vessels, resulting in the relative loss of his vision. The eye, while consisting of hundreds of highly developed genes, is supposedly functioning as "one unit." The eye has its own relative, independent complex self-motioned pro cesses; its own relative way of operating, moving, functioning and further evolving. But, it is not totally autonomous in its existence, behavior, deterioration, degeneration, and so forth. The eye outside of the body would not be a functioning eye! The eye without the nervous system, without the heart, without the metabolic system, without the general organic structure of the body, without the en vironment in which it lives and is, therefore, conditioned and in fluenced, while itself influencing and conditioning, does not maintain its absolute autonomy, and absolute material existence. The eye can only remain what it is within a well-integrated body, actively interacting with the environment. When my father had his diabetic conditions under control, there was a tremendous improvement in his vision. That definitely shows the organic in terrelated functioning of all organs of the body and their role in mutual conditioning and influencing in the processes of evolution.

Consider another example, showing the relativity of existence in gold, another part of material life. We have a chunk of gold in our possession. We are interested in finding out how much it weighs. We would weigh that piece of gold at sea level; and it would weigh, say, one ounce. If we take our scale and the chunk of gold to some high mountains and weigh our gold there, we would notice that the gold weighs less than one ounce. What is the explanation for that? Why would a chunk of gold be one ounce on the sea level, and less than one ounce at the mountain level? *The truth.* about it is *that the weight is not determined by the gold,* itself. It is determined in relation with the environment in which it is. The forces of gravitation and, therefore, the interaction of the gold and the earth molecules are much stronger as we get closer towards the center of the earth and relatively milder as we get farther from it. Therefore, our piece of gold would be heavier as it gets closer to the center of the earth, and lighter as it gets farther away from it. *Thus, we notice that the weight of the gold is modified, altered, mediated differently in different sets of condi tions.* If the greedy, but not so shrewd Isfahani gold merchant knew self-motioned atomic dialectical materialist logic, instead of penny pinching, by using faulty scales, he would buy gold at the mountain level and sell it at sea level. This way he could send all of his sons and daughters to American universities to be "educated" in the· tradition of John Locke mentality, while he fi nances them in a very honest and "truthful" way.

6 -RELATIVITY VS. ABSOLUTENESS OF TRUTH

What is the "true weight" of that piece of gold? Under two different circumstances, the chunk of gold had two different weights. Which is then the "true weight?" The truth is that truth is not a dead phenomenon; it is alive and in constant unfoldment. Truth is determined by the mutual interactions of self-motioned atomic entities, i.e., the interaction between gold molecules and earth molecules. *Truth is born out of actual material conditions, and its mode of existence keeps unfolding differently.* In the be ginning, we said truth must be the reflection of material existence. We already showed that material existence is mutually dependent for its continued existence and development. Self-motioned atoms, molecules, and macro-bodies are constantly overlapping, modify ing, altering one another, never in the same fashion. Therefore, truth, which is supposed to be a reflection of material life, must never reach an end. It must be conditional, that is, based upon the material conditions in which, it was obtained. It must rec-ognize mutual determination of all things. It must recognize change and evolution of all things. It must be born and reborn, never reaching an end. Because none of its births are exactly the same. The truth about your body when you are, say, twenty is never the same as that when you are twenty-seven. We are told by modern biologists that in a seven-year period, there is a complete rejuvenation of every cell in your body. The cells are never the same in two dif ferent periods.

J. Deitzgen's understanding of dynamic vs. dead truth is highly instructive: "Truths are valid only under certain condi tions and under certain conditions errors are true. It is a true knowledge that the sun is shining, providing that we assume a cloudless sky."[14] Well, you might ask: how can truth become error, and error become truth? These are supposed. to be two dif ferent things. How can the two overlap? Think about the way the weight of gold was determined differently under two different con ditions. Which one is the "true weight" of the gold? Which one is the unchanging, "absolutely true weight"; and which one is the absolutely false one? The point is ·that there is no absolute truth about anything. Truth is born differently under different condi tions. There is no absolute and once and for all truth. On this J. Deitzgen goes on to add:

The believers of absolute truth have adopted in their outlook the monotonous diagram of "good" men and "rational" in stitutions. For this reason, they oppose all human and historical institutions which do not fit into their pattern but which reality nevertheless, produces without regard to their brains. *Absolute truth is the· primary foundation of intolerance. Conversely, tolerance proceeds from the consciousness of relative validity of "eternal truth."* [10] [On the relative validity of general rea soning, Deitzgen concludes:] *Abstract or general reason, with eternal, absolute truths, is a phantom of ignorance, which binds the rights of individuality with crushing chains."*

In the following, Engels denies that truth is made up of an aggregate of well-polished, finished statements, which once estab lished becomes absolutely and eternally true:

Truth, the cognition of which is the business of philosophy, became in the hands of *Hegel no longer an aggregate of finished dogmatic statements, which once discovered had merely to be learned by heart.* Truth lay now in the process of cognition itself, in the long historical development of science, which mounts from lower to higher levels of knowl edge *without ever reaching by discovering so-called absolute truth, a point at which it can proceed no further and where it would have nothing more to do than to fold its* hands and *admire the absolute truth to which* it *had attained."*

On the un-attainability of absolute truth, Engels continues:

It is just the same with eternal truths. If mankind ever reached the stage at which it could only work with eternal truths, with conclusions of thought which possess sovereign validity and an unconditional claim to truth, it would then have reached the point where the infinity of the intellectual world, both in its actuality and in its potentiality, had been exhausted and this would mean that the famous miracle of infinite series, which has been counted, would have been performed.'"

On the relativity of truth, Engels holds:

Truth and error, like all concepts which are expressed in polar opposites, have absolute validity *only in an extremely limited field,* as we have just
seen, and as even Herr Duhring would realize if he had any acquaintance
with the first ele ments of dialectics, which deal precisely with *inadequacy
of all polar opposites.* As soon as we apply the *antithesis between truth and
error outside of that narrow field, which has referred to above,* it *becomes relative* and, therefore, unserviceable for exact scientific modes of expression;
and if *we attempt to apply it as absolutely valid outside that field we really find
ourselves beaten;* both poles of the antithesis become transformed into their
opposites- truth becomes error and error, truth.'"

On historical character of truth and the fact that human knowledge
would not reach a "perfect termination," Engels holds:

And what holds good for the realm of philosophic knowledge holds good
also for every other kind of knowledge and also for practical affairs. Just as
knowledge is unable to reach a perfected termination in a perfect, ideal condition of human ity, so is history unable to do so; a perfect society, a perfect
"state" are things which *can only exist in the imagination.* On the contrary,
all successive historical situations are only transitory stages in the endless
course of development of human society from lower to higher.

Each stage is necessary, therefore, justified for the time and conditions
to which it owes its origin. But in the newer and higher conditions, which
gradually develop in its own bosom, *each loses its validity and justification.
It must give way to a higher form which will also, in its turn, decay and per-
ish. just as the bourgeoisie, by large scale industry, com petition and the world
market* dissolves, in practice, all stable, time-honored institutions, *so this
dialectical* philosophy *dis solves all conceptions of final absolute truth* and of a
final absolute state of humanity corresponding to it. *For it, nothing is final,
absolute, and sacred.* It reveals the transitory, character of everything and

in everything; and nothing can endure before it expects the uninterrupted process of be coming and passing away of endless ascendency from the lower to the higher. And dialectical philosophy itself is nothing more than the mere reflection of this process in the thinking brain?0   [Still continuing on in-exhaustibility of relative truth, Engels continues:] . . . Knowledge, which has unconditional claim to truth, is realized in a number of relative errors; neither the one or the other *can be fully real ized except through an endless eternity of human exist ence."* [And finally, Engels contemptuously criticizes those who claim to be possessors and carriers of absolute truth:] When man is in possession of *the final and ultimate truth* and the only strictly scientific method, it is only natural that he should have a certain contempt for the rest of erring and un scientific humanity? •

1. V. Afanasyev, *Marxist Philosophy*, p. 171.
2. **Engels, *Dialectics of Nature*, p. 26.**
3. Ibid., p. 156
4. Ibid., p. 314.
5. Ibid., p. 269..
6. **Engels *Ant i-Duhring* (Moscow: Progress Publishers, 1969), p. 132.**
7. **Engels, *Outcome of German Philosophy*, pp. 14-15.**
8. **Engels, *Anti-Dulzring*, p. 31.**
9. Milac Capek, quoting J. Locke, *The Philosophical Impact of Con tempo-rary Physics* **(New York: Van Nostrand Co., Inc., 1961), p. 70**
10. **Maurice Cornfouth,   *Dialectlal Method,* quoting Engels (New York: International Publishers1970), p. 69.**
11. **Joseph Dietzgen,   *The Positive Outcome of German Philosophy* (Chicago:** Charles·Rerr Co., 1906), p. 107.
12. Ibid. p. 109.
13. **Engels, *German Philosophy* p, .44.**
14: *J. Dietzgen, _Po!itive OutCome of German Philosophy,* p. 114.
15. Ibid. pp. 158·159.
16. Ibid. p. 152.

17. Engels, *German Philosophy*, p. 11.
18. Howard Selsam and Harry Martel, *Reader in Marxist Philosophy*, quoting Engels, pp. 152-153.
19. Ibid., pp. 152-153.
20. Engels, *German Philosophy*, p. 12.
21. Reader in *Marxist Philosophy*, quoting Engels, p. 152.
22. Engels, *Anti-Duhring*, p. 41.